The **ARTHRITIS REVOLUTION**

Latest Research on Staying
Active Without Pain
Medication or Surgery

To Sam – my next GMAT success story

The ARTHRITIS REVOLUTION

Latest Research on Staying Active Without Pain Medication or Surgery

LOUIS G. PACK, D.P.M., M.S., F.A.C.F.A.S., D.A.B.F.A.S., F.A.C.R., F.A.A.F.A.S., P.C.

Founding Fellow, American College of Rheumatology
Former Clinical Instructor of Medicine, Emory University School of Medicine
Diplomate, American Board of Foot and Ankle Surgery
Fellow, American College Foot and Ankle Surgeons
Fellow, Academy of Ambulatory Foot and Ankle Surgeons
Faculty, Division of Enhanced Performance, United States Sports Academy
Adjunct Faculty, School of Orthotics and Prosthetics, Georgia Institute of Technology
Consultant, Injury Prevention & Enhanced Performance, U.S. navy SEAL's

The Arthritis Revolution
By. Dr. Louis G. Pack

Includes bibliographical references.
ISBN 978-1-4583-8974-9

Printed in the United States of America

Cover and Book Design by:
Becky Snowden
www.beckysnowden.com

DEDICATION

This book is dedicated to my dear mentor, the "Father of Modern Podiatric Medicine," Dr. M.D. Steinberg, who so inspired, guided and taught me the importance of continually searching for knowledge, so as to better care for those in need. A true genius and the epitome of integrity and professionalism, whose diagnostic acumen and treatment innovations I have yet to see surpassed. I was privileged indeed to have been his student for many years and one of his first medical residents.

*Make no mistake...I see further only because
I sat on the shoulders of this great man.*

TABLE OF CONTENTS

ACKNOWLEDGEMENTS

I wish to thank my children, Steve and Jeff, who inspired me to write this book, the many patients I have been privileged to treat and who taught me so very much, and those who permitted me to include their pictures, stories and testimonials. Friends, Ed Ralston and Fred Kaplan graciously guided me through the preparation of this manuscript. April Hayes helped with the organization and general editing. And Becky Snowden, my talented and patient graphic artist, who set the type, entered the pictures and designed our cover. Above all, many thanks to my dear wife and life's partner, Linda, who has always lovingly, offered support and painstakingly, performed the final editing on this manuscript

MEDICAL DISCLAIMER

The information contained in this book and the author's associated newsletters, websites, electronic and related materials are intended as educational aids only. Self assessments sections are designed to help you better determine if you have underlying problems that may make your arthritis worse. The recommendations and treatments presented are to be used as reference materials only. Pictures and case studies demonstrate certain positive results achieved over the years and are *not meant to imply or recommend in any way, self treatment. Nor are they intended as medical advice for individual conditions or treatment.* The material is presented to give you the tools you need to help make informed decisions about your health and to encourage you to take the appropriate steps in seeking professional advice for your arthritis. The information is *not* intended as a substitute for a medical examination, nor does it replace the need for services provided by medical professionals or independent determinations. Attempting to treat yourself may make your problems worse. The author and publisher expressly disclaim any responsibility for any adverse effects occurring as a result of self treatment from suggestions or information provided or implied herein.

EPIGRAPH

Known as the Semmelweis reflex, there is often an immediate rejection of new medical information, regardless of how sound or proven, simply because of narrow minded, deep-seated ideas. Yet, if we didn't radically change our thinking, medicine would never progress.

"Here's to the crazy ones........who see things differently. You can't ignore them. Because they change things. They push the human race forward. And while some may see them as the crazy ones, we see genius. Because the people, who are crazy enough to think they can change the world, are the ones who do."

Steve Jobs, co-founder and CEO of Apple,
former CEO Pixar Animation Studios

GETTING STARTED

Osteoarthritis, the most prevalent form of arthritis, is the leading cause of pain and disability in the United States. It plays a major role in our end-lessly increasing medical costs. And as life expectancy increases, so does the prevalence of this disease. Once osteoarthritis has developed, and pain medications no longer bring relief, joint replacement is often the only alter-native. But even this is not a permanent cure.

Until now, osteoarthritis has been thought of as a natural part of the aging process. Current thinking by most physicians, and therefore the public, is that other than losing a little weight and trying to stay active, there really isn't anything we can do to prevent or stop the onset of this dreaded dis-ease. Yet, the older I get, the more I appreciate just how important it is to remain active. Far too many of my friends and patients can't even play a round of golf or take a walk without pain. This doesn't paint a very pretty picture…and it's not.

For me, medicine is more than a vocation; it's a passion… my life's work. I love helping people and always have. I knew long before I became a physi-cian, that when I did, I would not be satisfied with simply *treating* someone; no, I wanted to be sure that I would know enough so that I could actually fix them…make them better. I have always felt a great responsibility and obli-gation to do this. Guess that's why my wife jokingly calls me, "Mr. Fix-it."

So I'm constantly searching for new answers for my patients. Now, after studying this subject for more than forty years, and successfully treating thousands of arthritis sufferers, of this I can be certain…

The current thinking that age is the root cause of osteoarthritis of the weight-bearing joints is incorrect.
And this, of course, makes the entire focus of treatment incorrect as well.

I will show you that although current thinking seems logical, some of it isn't. There is a difference between a *cause* and a *correlation*. The fact that osteoarthritis is seen in most elderly people doesn't necessarily mean it is caused by age. And this has great application in sports performance as well. So, by applying the very same principles I use to treat those with osteoarthritis, I will show that you may actually be able to *improve* your sports performance as you age!

The fact that I have been successfully doing this for most of my life, and continue to prove my premise daily on patients with this dreaded disease, may not be enough for the naysayers. And I can certainly understand and appreciate that. So I have added testimonials from some very prominent people, including some of the greatest athletes in the world...many of whom had problems just like yours. I will also share information from some important new studies at well-known institutions like the famed Mayo Clinic, which now substantiate my philosophy and clinical findings. In time, these findings will change how physicians think about and treat osteoarthritis. But medicine is very slow to react to new medical data, and you may need help now. That's why I wrote this book.

I will show you precisely why osteoarthritis develops, and what you can do immediately, to prevent or stop it from becoming worse...in many cases, without medicine or surgery.

This book is neither a comprehensive treatise nor a medical text on the various types of arthritis, nor is it a review of all the supplements, medications, or surgical procedures now available. Such information is readily accessible elsewhere.

As you read, please bear in mind that throughout the discussion, I focus only on *arthritic problems involving the weight-bearing joints...those joints of the feet, ankles, knees, hips, back, and neck.* Arthritis of other joints, such as the fingers and shoulders, is not addressed here, as it has other causes. Although my comments about preventing and treating arthritis and decreasing its symptoms refer primarily to osteoarthritis, they can also help in the treatment of other types of arthritis that affect the weight-bearing joints, such as rheumatoid arthritis.

Like most people, I find it frustrating to wade through writing I can't understand, so I've tried to make this book easy to read. I've simplified medical terms when possible and clarified ideas by providing understandable analogies.

A basic overview of osteoarthritis is presented first. Important current misconceptions are addressed, followed by a series of discussions on the real causes of osteoarthritis. The important role that the foot (the foundation of our entire skeletal system) plays in this disease is also presented. Step by step, you will see why and how osteoarthritis develops. And I think you too, will reach the same point of logical conclusion that I have. Then you will learn exactly what you can do to help eliminate your symptoms and stay more active. For those of you who wish to take things a step further, there is also a section on increasing your performance as you age.

Each chapter's major points are listed at the beginning of the chapter to help you recognize and remember them. I've included many clinical pictures to explain the concepts; concepts that I will repeat throughout this book, because your success in eliminating symptoms depends upon your clear understanding of them.

Although I will show you how to recognize the real causes of your osteoarthritis, *this book is not intended to advise you on self-treatment.* Rather, by understanding the information, you should be better able to take a proactive role in getting the treatment you need.

Whether you're reading this book to prevent osteoarthritis, lessen your symptoms, or improve your sports performance, I trust you will find this cutting-edge information helpful…information that is being increasingly validated every day. By following my suggestions, most readers should see some improvement. In many instances, these simple yet powerful techniques can help immediately. And for some of you, this information may just change your life as it has for countless others I've treated over the years.

The concept that age is the primary cause of osteoarthritis must change for you to get the care you need. This change should begin with pediatricians who are the first line of defense in preventing often easily fixable deformities from becoming serious ones, and be continued by primary care and family practice physicians, internists, rheumatologists and others who are involved in your care.

I sincerely hope you find this book both helpful and enjoyable. Thank you for reading my life's work.

Dr. Lou Pack

www.drloupack.com

CHAPTER 1

Osteoarthritis
an OVERVIEW

"Chronic knee pain due to my osteoarthritis was always a part of my life, or at least, so I thought. Even standing still at a party caused my knees to hurt. After working with Dr. Pack, that pain is all gone, and the quality of life has never been better. I'm hooked. I can even enjoy playing golf again without pain."

Steve Melnyk

Winner of 1969 U.S. Amateur and 1971 British Amateur Golf Tournaments, University of Florida Hall of Fame, Florida Sports Hall of Fame, and the Georgia Golf Hall of Fame; former golf analyst, CBS and ABC Sports

- Osteoarthritis is the leading cause of pain and disability in the United States.

- Age, increased weight, and genetics are currently thought to be the primary predisposing factors for this disease.

- Pain medications, supplements and surgery have a place in treating osteoarthritis, but do not address and therefore, eliminate the underlying cause.

HOW BIG A PROBLEM *IS* ARTHRITIS?

Arthritis affects 70 million Americans or about one in every three adults [1]. This disease accounts for 44 million outpatient visits and three-quarters of a million hospitalizations a year. According to the National Arthritis Data Work Group, the estimated cost to society, including lost work productivity, is $83 billion a year [2].

Osteoarthritis is the most common form of arthritis. Despite the introduction of many new drugs and advanced surgical procedures, misunderstandings regarding the true cause of this disease have resulted in medicine's inability to prevent or stop it. This has led to an epidemic number of osteoarthritic patients.

In the next chapter I will share new, important information that will begin to help you... not just with osteoarthritis...but with other forms of arthritis as well. But first, let us define our subject matter and briefly discuss today's accepted medical data...data that, as you will soon see... is not entirely ac-curate regarding the causes and treatment of osteoarthritis.

ARTHRITIS AND OSTEOARTHRITIS

DEFINITIONS

Arthritis is not a New Age disease. To the contrary, evidence of arthritis has been found in dinosaurs and Egyptian mummies. As much as 70% of the population of ancient Rome had arthritis; one of the reasons their many public baths were built. Often these, like the ones in Bath, England, utilized natural hot springs, because of the soothing effect the heat had on reducing inflammation and relaxing tight muscles.

There are more than one hundred different types of arthritis. In some cases the disease is localized. In others, it may be classified as systemic, sometimes affecting the eyes, lungs, heart, kidneys and other parts of the body as well as the joints. Unfortunately, most types have no known cause.

Arthritis is primarily thought to be a disease of the elderly, most commonly causing inflammation of the joints. Some forms, however, such as juvenile rheumatoid arthritis, can affect young children. Others, such as fibromyalgia, primarily affect muscles without signs of joint inflammation. Diseases such as polymyalgia rheumatica can cause muscle soreness without joint involve-ment. Over the years, classification terminology has changed and now, more accurately labels such conditions as "arthritis related."

Sometimes the term *rheumatism* is also used. Derived from the Greek word *rheumatismos*, it refers to the swelling that occurs in some types of arthritis.

Unlike many arthritic conditions that have been painstakingly categorized according to more clearly defined diagnostic criteria, is a catch-all term that refers to vague muscular aches and pains and joint soreness.

As you can see, due to the frequent use of inaccurate terminology, this subject can be confusing...even to physicians

BASIC FACTS ABOUT OSTEOARTHRITIS

Because of destructive joint changes, osteoarthritis is often referred to as a wear-and-tear disease. It is the most prevalent form of arthritis... more common than all other forms combined. Generally associated with age, osteoarthritis is thought to be inevitable...something we will all get along with gray hair and wrinkles.

TYPES OF OSTEOARTHRITIS

Two types of osteoarthritis have been identified: primary and secondary. Primary osteoarthritis is thought to be mainly due to age (over 45) and excessive weight. Because it seems to run in families, it is presumed to have a genetic component. More than half of the cases of osteoarthritis affect the knees and occur more often in females. All of the weight-bearing joints are prime targets, as are the joints of the fingers.

As the name implies, secondary arthritis is due to an underlying condition and typically occurs in people under 45 years of age. Trauma is thought to be the main cause, but infections, excessive flexibility, thyroid abnormalities, gout and other medical conditions can also result in this type of arthritis.

This book focuses on primary osteoarthritis of the weight-bearing joints (feet, ankles, knees, hips, back, and neck), which is currently thought to be due to the predisposing factors listed in the following section. What I believe to be the real causes of osteoarthritis of the weight-bearing joints and their subsequent treatment, do not apply to the non weight-bearing joints such as those of the fingers and elbows.

PREDISPOSING FACTORS

1. AGE

According to the National Institutes of Health, osteoarthritis is thought to be related principally to old age and is both normal and unavoidable [3]. Indeed, currently, if there is one single factor consistently associated with the cause of osteoarthritis, it is age. Live long enough...and you can count on having it!

2. WEIGHT

Because of the additional mechanical stress it places on cartilage, obesity is ranked second only to age as a risk factor for osteoarthritis, especially osteoarthritis of the knees. The Arthritis Foundation advises that for every extra pound you gain, three pounds of pressure are added to your knees and the pressure on your hips is increased six times.

The famous Framingham Study, which began in Massachusetts in 1948, showed that among women, the heaviest had over twice the risk of knee arthritis as the lightest [4], while an English research team found that the risks of arthritis were as much as 13 times greater in heavy people. They concluded that by losing 15 pounds, 24 percent of the knee operations for arthritis could be avoided [5].

So, losing even a little weight can have a profound effect on reducing symptoms and slowing the arthritic process. As a matter of fact, according to the Arthritis Foundation [6], for some women, the risk of osteoarthritis of the knee can be reduced by half with the loss of a mere 11 pounds.

3. GENETICS

The genetic correlation for this disease is not as great as that for rheumatoid arthritis and some other joint disorders. Although thought to play some role in osteoarthritis, no specific "arthritis gene" has been identified. However, a genetic tendency does seem to run in certain families, such as those whose members have osteoarthritis in their hands. It has also been found that siblings of people, who have had joint replacement surgery for osteoarthritis, are about four times more likely to require similar surgery [7].

PATHOLOGY AND SYMPTOMS

Although osteoarthritis affects many aspects of the joint, including the capsule (joint covering) and its muscular attachments (tendons), cartilage is the focal point of involvement. Primarily designed as a protective covering to reduce joint friction, this important structure becomes eroded. And much like the callus tissue that develops on our skin in response to increased friction, in osteoarthritis, extra abnormally hardened bone is protectively produced to replace the lost cartilage, and to prevent further bony erosion.

Initially, joint pain can be very mild and transient, usually consisting of a dull ache after use, then disappearing with rest. Stiffness, especially in the early morning, is a cardinal feature of early osteoarthritis and may even precede the onset of pain. Although this can be quite disconcerting, it usually lasts only a few minutes, in contrast to the hours of "loosening up" required with rheumatoid arthritis.

As the disease progresses, joint spaces decrease, causing bone to rub against bone (See Figures 1A, B, C, and D). Roughened, fragmented cartilage and bone can break off into the joint, producing crepitus...a crackling sound coming from the affected joint. Bone spurs and cysts are also common, and in severe cases, large amounts of inflamed joint fluid, which becomes thinner and less lubricating, can cause significant swelling.

Eventually, all of these changes combine to cause extreme pain, even at rest. This is accompanied by a "locking" or stiffening of the joint which is a classic symptom of this disease. In response to the pain, and as a protective mechanism to avoid additional joint damage, muscle spasm occurs to guard against further motion. Unfortunately, this can cause even greater discomfort.

CURRENT TREATMENT
WHAT WE'RE DOING ISN'T WORKING

Although great advances have been made in many aspects of medicine, arthritis experts Doctors David Hunter and Grace Lo, summarize the current situation as follows:

"Knee OA is the most prevalent cause of mobility dependency and disability. Despite growing concern, *OA remains a poorly understood disease, and*

Figures 1A and B are lateral (side) views of an ankle joint as seen on X-rays. Figure 1A, shows a normal joint space. Figure 1B, shows partial obliteration of the joint space as seen in osteoarthritis. This indicates that the cartilage, the protective joint covering that normally allows a gliding, friction - free motion, is gone. Movement is then very much restricted and painful because bone is rubbing against bone... commonly referred to as "bone on bone" arthritis.

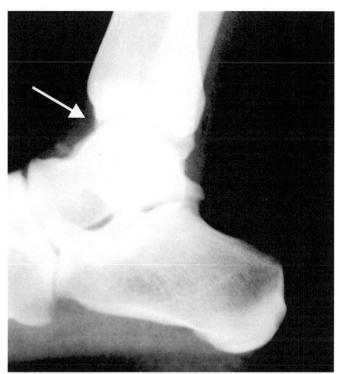

FIGURE 1A

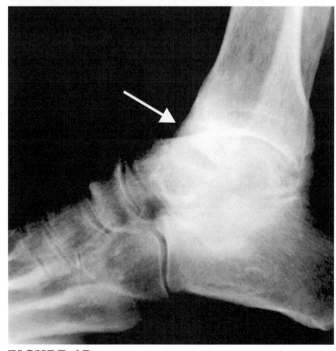

FIGURE 1B

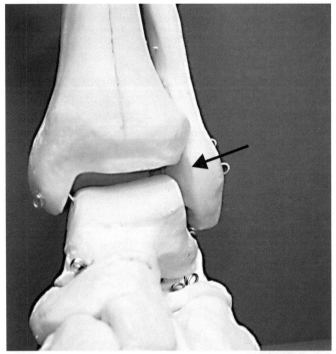

FIGURE 1C

Front skeletal views
of an ankle joint.
Figure 1C, shows a
normal joint space,
while partial
obliteration and
lateral compression
of this space is
depicted in
Figure 1D.

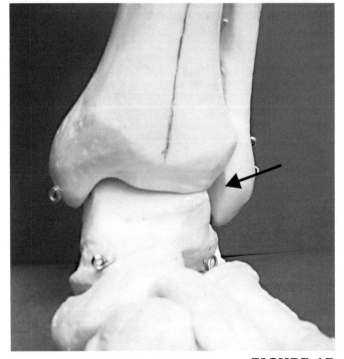

FIGURE 1D

recent doubts about the safety of several commonly prescribed OA medica-
tions have served to highlight deficiencies in the traditional medical approach
to management. Current clinical management for OA often is limited to an-
algesic medication and cautious waiting...until referral for total joint replace-
ment becomes necessary." [8, p. 689]

The term, *"cautious waiting,"* just mentioned above, is used by the medical
community to mean, "delaying definitive treatment while the disease slowly
progresses." *Far more than just a term, this is unfortunately, the standard of*
care the world over; certainly not because physicians aren't concerned or take
this disease lightly, but frankly, because they do not know what else to do.

According to the Arthritis Foundation, one in every two people is now at risk
of developing osteoarthritis of the knee, many of whom will eventually need
joint replacement surgery. Indeed, for many arthritis sufferers, knee and hip
joint replacements have become inevitable. Millions of people have undergone these procedures and the numbers are increasing every year. If you
haven't had a joint replacement, the chances are you certainly know someone who has. And remember...even when successful... these procedures
do nothing to stop a person's disease from getting worse, since they don't
alleviate the factors that caused it to begin with.

Because of the current medical opinion that there is a direct correlation between osteoarthritis and aging and that there is no known method of preventing or curing this disease, more and more of your friends and mine are
becoming afflicted. It has been estimated that if something doesn't change...
and soon...by 2030, 41 million older adults in the United States will have this
painful, disabling disease [9, p. 3].

THE TYPICAL SCENARIO

Recently, you've noticed that your knee has begun to hurt every so often or
that it gets stiff after you've been sitting for a while. After weeks or months
of waiting for the pain to "go away on its own," you visit your primary
doctor, who, before even completing the examination, says, "You probably
have some arthritis. We'll get X-rays and take a closer look." Sure enough,
the films show some joint space narrowing, bony proliferation, and other
changes indicative of osteoarthritis. Without fanfare, you've suddenly joined

one of the largest clubs in the world. But somehow, hearing, "It's a normal part of the aging process," and "We're all going to get it," doesn't bring you much comfort.

Now begins the medication regimen. Since no known cure is recognized, the goal of therapy is to minimize your discomfort so you can stay active. Pain medications, anti-inflammatory drugs and supplements are typically prescribed. As your joint pain becomes more than incidental and medication no longer works as well or for as long as it once did, you get a consultation from an orthopedic surgeon.

If you haven't had your problem very long, you'll typically be told, "Come back when it's bad enough to replace your joint."   This is cautious waiting, and that's exactly what a great many people do. According to the National Center for Health Statistics, in 2004 there were over half a million hip and knee replacements in the United States [10]. The National Institutes of Health (NIH) reports that more than 300,000 total knee replacements are performed each year in the United States in the Medicare population alone [3], while in a recent study conducted in Pennsylvania by the Pennsylvania Health Care Cost Containment Council, researchers showed that between 1993 and 2002, the number of knee replacements increased more than 70 percent, and hip replacements more than 48 percent [11, p. 1].

At this point, having exhausted their options, many patients are left "hanging."   Despite their belief that aging and overweight are primary predisposing factors, physicians really don't know why you have developed osteoarthritis. So, other than continuing to take your pain pills, which have risks of serious side effects that increase with higher dosages and longer periods of use, you're supposed to keep doing what you've been doing... which may have caused the problem in the first place...until you can no longer stand it! As a matter of fact, you're encouraged to use those painful joints to keep them healthy...tough to do when they hurt even at rest.

Certainly, there should be a better alternative than "cautious waiting" until you need to replace your arthritic joint. When that time comes, your problems, as you will soon see, are usually far from being permanently corrected, because you're actually starting the process all over again! And

I do mean again…because that shiny new joint will usually need to be replaced, too. *Again, this unfortunate scenario is taking place every day in physicians' offices all over the world.*

Now, if one of the tires on your car was wearing out faster than the others, would it make sense to "cautiously wait" until it wore out completely and then simply replace it and begin the process again? I don't think so! You would certainly want to find out why this was happening and fix the cause. And if your mechanic didn't have the answer, you would find another mechanic who did.

Yet, people who have arthritic joints must accept this as the norm, because unlike finding another mechanic who could fix your abnormally wearing tires, current medical thinking limits what most physicians can offer you…which again, is nothing more than a recommendation to continue using your knee or hip until the pain and damage necessitate its replacement.

TAKING A CLOSER LOOK

PAIN MEDICATIONS

While no one likes pain, we're designed to have it. Pain warns us of impending danger and alerts us to the fact that if we keep doing the activity that caused it, not only will the pain worsen, so will the damage. That's not to say that pain medication should not be taken after a surgical procedure, accident or in other appropriate situations. But taking it long term for arthritic pain does nothing to eliminate the cause, and may allow you to continue an activity that your body is clearly telling you not to do, and result in greater damage.

Pfizer Inc., for example, a well recognized drug company, had to recently remove an experimental drug it was testing for this very reason. Tanezumab, eliminated so much pain, that arthritic patients were causing permanent damage to their joints.

The same is true in sports. Many people pride themselves on being tough and persist in an activity despite their pain. This is especially true of runners, who often say with pride, "I ran through the pain." Not a very good idea.

By doing so, they cause far more damage, thus requiring a longer recovery period than if they had heeded this important, built-in warning sign. I know this first hand, as I have been a runner since 1972.

FURTHER THOUGHTS ON MEDICATIONS

My own experience with medications goes back a long time. My father was a pharmacist and had his own drugstore. If I wanted to see Dad, I always knew where to find him. I started working there as a young child and by the time I was a teenager, I had a pretty good understanding of drugs.

After attending pharmacy school and studying the uses and side effects of countless medications, I learned one lesson above all *else...the less medication you take and as a physician prescribe, the better.* Remember, anything you put into your body that isn't supposed to be there has risks which should be carefully weighed. *Even if the risk factor for a negative outcome is only 1 in 1,000, it's 100 percent when it happens to you!*

These risks are obviously greater with prescription drugs. The fact remains, if they were as safe as we're often led to believe, we wouldn't need a signed document ...a prescription... that only a licensed physician can write, to purchase them. And the drug companies that manufacture them and the physicians who prescribe them wouldn't be liable for the problems that they may cause.

In some circumstances, drugs prescribed to eliminate a problem can actually cause it. This is true for example, of some of the mediations used to prevent fractures in those with osteoporosis, as well as some used for the treatment of certain kinds of cancer. And as you will see below, this is true of some of the medications you may be taking for your arthritis. But there can be serious complications with over-the-counter, nonprescription medications as well. This is validated by an overwhelming number of studies.

Four types of pain medications are currently used to treat osteoarthritis: acetaminophens; NSAIDs (nonsteroidal anti-inflammatory drugs); a variant of these, the COX-2 NSAIDs; and synthetic steroids. Acetaminophens (such as Tylenol and Liquiprin) are over-the-counter medications that lower pain and fever. The NSAIDs also reduce inflammation. These include such over-the-counter preparations as aspirin, Advil, Motrin, Aleve, Naprosyn, and their prescribed versions, Voltaren, Indocin, Mobic, Naprosyn, and Feldene. The COX-2 inhibitors reduce inflammation also, and include medications like Vioxx and Celebrex. Both the COX-2 NSAIDs and the synthetic steroids (such as cortisone and prednisone) require prescriptions.

Because they are generally safer, acetaminophens are usually prescribed first for those with osteoarthritis, especially if no inflammation is present. But studies have shown that less than 2/5 of patients get pain relief even at the maximum dosage [12]. The possibility of serious side effects is also of great concern...even with these drugs. As a matter of fact, from 1993 to 1999 there were a reported 56,680 emergency room visits, 26,000 hospitalizations and about 450 deaths per year attributable to these "safer" medications [9, p. 124].

The other three categories of drugs...NSAIDs, COX-2 NSAIDS and synthetic steroids, are more potent and therefore, have even greater potential consequences. Currently, 13 million Americans take NSAIDs, which in this country alone have caused 16,500 deaths and 103,000 hospitalizations a year [13, 14]. Further, long-term use of steroids can cause diabetes, cataracts, thinning of the bones with subsequent fractures and many other serious complications. As quoted earlier, "Recent doubts about the safety of several commonly prescribed osteoarthritis medications have served to highlight deficiencies in the traditional medical approach to management." [8, p. 689].

If the risk of serious side effects wasn't enough cause for concern, *some studies indicate that not only may these drugs be ineffective...they can actually make your arthritis worse* [9, p. 122]! Evidence suggests that the COX-2 drugs, for example, can cause increased cartilage loss and "may impair the natural healing response in cartilage" [p. 137]. *And because painful symptoms can actually improve while this is happening, patients may be totally unaware of the increased damage that's occurring.* Furthermore, rarely do any of these drugs totally relieve symptoms long term. In a 2008 journal article, published in the Rheumatic Disease Clinics of North America, researchers reported that, "In clinical trials, all the most commonly used NSAIDs have performed comparably, with subjects reporting approximately 30% reduction in pain, and only 15% improvement in functional tolerance" [15, p. 756].

Because of the problems with these medications, safer, nonprescription medications have gained popularity. Supplements such as glucosamine, chondroitin sulfate, and ASU (avocado soybean unsaponifiables) are becoming more widely used in the treatment of osteoarthritis. Favorable extensive research and clinical responses, coupled with a lack of side effects, have certainly helped to increase their use. Working synergistically to decrease cartilage breakdown while helping to produce new cartilage, these substances claim to treat osteoarthritis on the cellular level, far better than simply trying to decrease pain or inflammation.

But no definitive conclusion has been made regarding the efficacy of these products. Critics claim that when benefits have been found, they tended to be trials that were industry-funded and not those without commercial sponsorship. Because of this, as of the writing of this book, the American Rheumatology Association has not formally endorsed taking these medications. It is therefore suggested that you discuss these supplements with your physician before taking them.

Remember...nothing is a cure-all. These products don't work for everyone and again, have not been conclusively proven effective. And although they may be a far better form of "air for a leaky tire," like pain medications, anti-inflammatory drugs, or even surgery... these supplements do nothing to eliminate the cause of your osteoarthritis.

SURGERY

For most of my professional career, I practiced surgery. And I can tell you that replacing an arthritic joint without understanding the mechanisms that caused that joint to deteriorate, is incomplete treatment at best...as incomplete and unacceptable as thinking that our only alternative to osteoarthritis is to medicate it away, or simply replace it, and let the process begin all over again.

Because it's far less invasive and offers a much quicker recovery time, arthroscopic surgery has become increasingly popular. Although it has proved to be a successful technique in many other kinds of joint surgery, this approach offers very little for a severely arthritic knee. A recent study, published in the prestigious *New England Journal of Medicine*, found that, "arthroscopic surgery for osteoarthritis of the knee provides no additional benefit to optimized physical therapy and medical therapy" [16].

So, when all else fails, a total joint replacement is the procedure of choice for arthritis of the knee and hip. Patients, who ultimately have this surgery, usually feel that their problem is finally solved. That is certainly understandable, but unfortunately, often far from reality.

Both knee and hip joint replacement surgeries are generally considered safe. However, as with any major surgery, there are the significant risks of infection, pulmonary emboli, and other complications. *And even when considered successful, a new knee will not generally have the full, complete range of motion that your original knee once had. So you will not have a "normal knee," even in a best case scenario.*

In addition, researchers have concluded that the more active you are, the sooner your replacement will wear out. Indeed, according to a recent article in the journal *Rheumatic Disease Clinics of North America*, "…there are significant issues concerning the long-term durability of total knee arthroplasty in the younger population (less than 55 years of age)," and, in addition, "One long-term study of more than 10,000 total knee surgeries demonstrated that the failure of total knee replacement at 10 years, was approximately three times higher in the population under the age 55 years, than in the population over age 70 years." [17, p. 815] That's why joint replacements are generally indicated for older patients who are less active.

Some researchers project that "if current practice patterns continue, the number of total knee arthroplasties [replacements] performed annually in the United States will increase 525% by 2030, and annual costs for knee and hip replacement will reach $62.5 billion by 2015" [18]. Is it any wonder that I say," *What we're doing simply isn't working?"*

Your surgeon may have been able to get you on the road again by putting a "run-flat" tire on your car, but those replacement tires may have risks associated with them and will never give you the performance or the same comfortable ride that your original ones did. Similar risks and complications hold true for total hip replacement surgery, with the exception that these last a little longer in the elderly population. But forty percent of hip replacements are performed on people younger than 55 [19]. Therefore, these will typically wear out much faster because of greater activity.

The most prevalent early complication of total hip joint replacement is joint dislocation. Later, inflammatory reactions caused by small particles wearing off the implants and becoming absorbed by the surrounding tissues are commonly seen. Inflammation can cause the implants to loosen, sometimes necessitating additional surgery. Leg-length discrepancies are also very common...far more common than reported. At times, differences of one inch or more in leg-lengths can be seen following hip replacement surgery. Although they have neither been as severe nor as frequent, I have seen these same problems with knee replacements.

You should also know that like drugs, the jury is always out regarding the safety of these procedures. Often, the latest materials and techniques are hailed as "new and improved," yet later they're sometimes found to be not only ineffective, but dangerous. After selling more than 56,000 of them, Zimmer Holdings, Inc., recently had to remove its new hip implant from the market because of major complications [20]

On March 4, 2010, the New York Times reported that "some of the nation's leading orthopedic surgeons have reduced or stopped use of 'metal-on-metal hip implants'. This was amid concerns that the devices were creating large amounts of metallic debris that were being absorbed into the patient's bodies."

The Wall Street Journal reported on August 28, 2010, that DePuy, a division of Johnson and Johnson, issued a recall of 93,000 of its hip implants because of pain, difficulty in walking and other complications. The recall was issued some two years after complaints were received by the Food and Drug Administration regarding the failed implants.

All implants create some debris as one part of the implant wears on the other. But often in the case of metal-on-metal implants, the excessive amount and type of particles (metallic) not only result in pain, but cause the actual death of tissue in the hip joint, as well as loss of bone in the area. The corrective surgery that may be necessary is usually quite difficult and complications can cause permanent damage.

It is important to realize that these problems can occur because of poor implant design, improper surgical techniques, or simply the increased frictional forces on the adjacent surfaces of the implant itself. Regardless of the later…anything that is improperly aligned will wear out more quickly.

Now, that's not to say that current care is all doom and gloom. Pain medications can be helpful, and I'm certainly glad they're available when needed. Supplements like glucosamine have helped many people with arthritis. Knee and hip replacements have enabled millions of people to resume active lifestyles. But when you consider some of the risks I've mentioned, and the fact that none of these therapies prevent or cure osteoarthritis…I think you would agree that the best treatment is the one you don't need at all…and that's especially true of pain medications and surgery.

Medicine for all its wonderful advances, is still for the most part a fix-it profession. Unfortunately, we place far too much emphasis on patching things up once the damage is done, instead of on finding and eliminating the cause. Let's face it, most of your visits to a doctor are because of a problem you already have, not to prevent one. Generally, what doctors do is focus on responding to those problems.

Now, all this can change. As I'll show, there is much you can do to prevent osteoarthritis or stop its progression. But to do so, you must begin to understand its real cause.

" All I had to offer osteoarthritic patients was pain medications. Now, thanks to Dr. Pack, I am so excited to be able to offer them something else that can really help them! "

ALYCE OLIVER, MD, FACR
Assistant Professor of Medicine, Department of Rheumatology,
Medical College of Georgia

CHAPTER 2

Questioning
WHAT WE HAVE LEARNED
CURRENT MISCONCEPTIONS

"*Dear Dr. Pack,*

I would like to offer you an unsolicited testimonial. It was almost impossible to exercise and lead a healthy lifestyle before you treated me because of severe disabling pain. My favorite past time, golf, was actually shattered. Now, besides total pain relief, I am able to jog two miles every day…. something I never thought I would be able to do again… without pain. At 70, the balance you have given me has made my golf game miraculously return. And amazingly enough, I am striking the ball with much better precision and have actually added distance to all my clubs. You cannot possibly appreciate the full extent of the help you have given me and how very much it has changed my life."

Sincerely,
Albert G. Applin, Ph.D.
Former Dean of Academic Affairs and Instructional Design,
United States Sports Academy

> • With new information, the practice of medicine changes over time. If physicians did not periodically change their thinking, medicine would never progress.
>
> • A great void in traditional medicine is the lack of acceptance and integration of structural abnormalities as they relate to the cause and treatment of osteoarthritis.
>
> • There are many misconceptions regarding osteoarthritis that are directly responsible for the void in treatment many arthritic sufferers experience.

THE ROAD TO ANSWERS

As we acquire knowledge, we occasionally realize that some of the things we have been taught aren't exactly correct. George Washington's cherry tree story is more fable than fact, and Christopher Columbus was not the first to discover America. The same is true of medical information. *As physicians, we realize from time to time that some of the "facts" we once believed are no longer valid, so we have to change the way we think and practice. This is essential in order for medicine to progress.*

Medical thinking has undergone many changes…even during my own lifetime. As a child, I was told to wear only white socks, because I had athlete's foot; a ridiculous idea today. Perhaps, like many of you, I was never allowed to swim less than an hour after eating, lest I suffer cramps in the water and drown…a dictum considered questionable since 1961 and never proved [1]. A "good breakfast" once consisted of whole milk, bacon, eggs, and toasted Wonder Bread to 'build your body 12 ways." If you're Southern, you probably added pancakes and grits. When you "cleaned your plate," your mom may have allowed you to top off that "healthy" breakfast with a Hostess Twinkie, Snowball, or a handful of Oreos…and maybe some chocolate milk to wash it all down. Today we are told that a "healthy" breakfast like that might cause us to have a coronary before lunch! Of course then, a cholesterol level of 300 mg/ml was considered normal.

Long ago, physicians once thought that bleeding patients was helpful in treating many ailments, and that washing hands and instruments wasn't necessary when doing surgery. Initially, cell phones were thought to be totally safe. Recently, we have become concerned that our children will get brain tumors from using them too much.

Now the time has come for medicine's thinking regarding osteoarthritis to change too, because the current thinking is outdated [2]. This is critically important, because changing the thinking about what causes a disease changes the way we treat it, and that changes the end result...something that is really needed and directly affects you.

SPECIALIZATION AND OSTEOARTHRITIS

During the last century, tremendous advances in medicine resulted in an era of specialization. Today, we not only have specialists, but subspecialists... professionals who know more and more, but in a limited area. Now a patient has the option of not just seeing an ophthalmologist for an eye problem, but among others, a corneal or retinal specialist.

That's great...if you see the right doctor for your particular problem. But if you don't happen to choose correctly, you can become one of the "wandering wounded," going from physician to physician looking for answers. That's because these otherwise, well-trained doctors, typically view a problem from the point of view of their own specialty. This is one of the reasons I think that osteoarthritis is considered to be a disease of the elderly. The early signs of osteoarthritis remain, for the most part, unrecognized and untreated, first by pediatricians. So physicians seeing older patients can only assume that age is a primary factor, since most of their elderly patients have this disease.

One great void in traditional medicine has been the lack of acceptance and integration of structural abnormalities...like a longer leg or flattened foot... as they relate to osteoarthritis. Since we are structural beings and much of how we feel and function is dependent upon the integrity of our skeletal system, it is surprising that this is not a major consideration when patients are evaluated. Instead, this area has been virtually relegated to the research of biomechanists, who study the relationship of mechanical laws to human motion and to the adjustments and manipulations provided by chiropractors and physical therapists. As a whole, mainstream medicine, which includes general practitioners, internists, orthopedists and rheumatologists, has not embraced the important concept of biomechanics or structural analysis. And until it does...patients with osteoarthritis will not receive the complete treatment that they need.

My perspective is different because I have had the opportunity to increase my scope of knowledge by studying with specialists in a number of different areas, including foot and ankle surgery, biomechanics, sports medicine, diabetes and arthritis. Quite frankly, like many physicians, I failed initially to realize the importance of structure and function as they relate to osteoarthritis, believing that this area had little significance in "real" medicine.

THE JOURNEY THAT CHANGED *MY* THINKING

Early on, my primary area of interest was the pedal (foot) manifestations of systemic diseases. I focused on how various diseases such as arthritis and diabetes affected the foot. I was trained by Dr. Marvin D. Steinberg, who was thought to be the greatest mind in this specialized area of medicine. Biomechanics (mechanical principles applied to human function) however, was totally absent from his focus and therefore from the curriculum.

Later, I became interested in surgery and was accepted to a program in San Francisco. Here the emphasis was on biomechanics and advanced surgical techniques for foot and ankle pathology. No longer was I to think in terms of diseases. The entire focus of my training was now on understanding human mechanical function and the surgical correction of problems that we otherwise could not fix. Physicians on the West Coast were seeing exactly the same types of patients that I had seen in the East, yet they were diagnosing and treating totally different conditions. Once again...physicians typically see only what they are trained to see. You may have heard the cliché, "If all you have is a hammer...everything is a nail."

Initially, I wasn't a believer. As a matter of fact, because of my prior training and the great esteem in which I held Dr. Steinberg, I was very skeptical indeed. However, the Bay Area in the early 1970s was the center of biomechanics and the sport of running, which was rapidly gaining in popularity. Before long, I too began to run, finding it a great way to reduce stress and increase my focus. But after my first Bay to Breakers race, I developed a stress fracture in my upper fibula, which is the outer leg bone below the knee. Despite all of my medical training and quite a long period of rest, I was unable to resume running without pain.

With great reluctance, I consulted a colleague who was an expert in biomechanics. After examining me and finding structural abnormalities that he believed were adding stress to my injured leg, he made me a pair of custom running foot orthotics (shoe inserts). Although initially convinced that I had wasted my time, the very first day I put them in my running shoes I was able to run five miles without any pain whatsoever.

The effect that had on my thinking was far greater than the pain relief I suddenly experienced. For the first time, I began to appreciate the powerful role that abnormal mechanics plays in human function.

Runners will generally suffer injuries because of the undue stress they put on their systems. In the 1970s before the development of the sophisticated technology of today's running shoes, better training methods, and other improvements, runners had more problems, and more often. A prime example was a condition called *runner's knee.* This is a term coined by the late prominent cardiologist, runner, and author, Dr. George Sheehan, who was among the first to emphasize the importance of proper foot positioning in eliminating this condition.

Applying what I had learned and experienced myself to other runners, I was continually amazed by what I was seeing…*that optimizing foot position… rather than focusing on the knee itself…was often responsible for the dramatic elimination of knee pain.* As you will see in Chapter 7, many medical studies substantiate this.

Other things continued to reinforce this new trend in my thinking. During my surgical training, I operated on young patients who had similar osteoarthritic joint changes that I had been taught occurred predominately in the elderly. At that time, we used plastic implants instead of the sophisticated metal ones used today to replace damaged joints in the foot. Yet over time, regardless of our techniques or expertise, some of these also had to be replaced due to wear and tear.

This concerned me a great deal. I was very close to my grandfather and held him in great esteem. He was a craftsman that could build most anything, and took great pride in them standing the test of time. Indeed, if he put a chinning bar up in my doorway and an earthquake hit, it might be the only thing left standing. His company was appropriately named, "Everlast." Each time I had to replace a joint I had previously done, I could hear his disappointment in the longevity of my work.

As a car enthusiast, I appreciated the importance of optimizing wheel alignment in making tires last longer. I knew there had to be something that could be done to prevent this type of repetitive surgery. Once I realized that I wasn't really addressing the underlying cause of the problem...why the replacement joints were wearing out...I began to experiment with different means of decreasing the stress on these areas by improving structural alignment. *As a result, I was often able to prevent the surgery or the need for repetitive joint replacements.* Indeed, I've had patients who have been able to run on those now archaic plastic joints for nearly 30 years and have never had them replaced.

After finishing my surgical training, I became interested in diabetes and received an appointment as a Clinical Instructor of Medicine, in the Division of Endocrinology, in the Diabetic Foot Clinic, at the Emory University School of Medicine. Among other things, diabetes affects the feet, and can cause serious ulcerations and subsequent amputations. Despite the fact that excellent physicians did their best to control the blood sugar levels of our patients, in some cases amputation was still inevitable.

I realized that the term *diabetic ulceration* was something of a misnomer. Although diabetics can certainly develop ulcers due to their inherent circulatory problems, they usually do not develop them on the bottoms of their feet just because they have diabetes. There is something else at work.

We all have structural abnormalities that result in abnormal frictional forces. Normally, a healthy patient responds to these forces by forming a corn or a callus on their foot, which is a protective area of hardened skin. In the case of the diabetic, the skin breaks down and becomes infected because it is far more fragile. So in many instances, these ulcers aren't really *diabetic...* that is, *primarily caused by* diabetes... but rather *precipitated by it;* i.e., they are *mechanically induced* ulcers that are made worse by the disease.

[1]Also known as patellofemoral pain syndrome, runner's knee is a condition in which pain occurs in and around the knee that is often seen in people who are engaged in active sports.

After instituting medical and surgical techniques to treat the infections, I once again found that by altering the stress forces on the feet of diabetic patients, I could prevent further tissue breakdown and often avoid leg amputations. For a number of years, I lectured and published on limb salvage in these types of cases [3, 4, 5].

It is interesting to note that in both diabetes, and as you will soon see...osteoarthritis... the underlying pathology is abnormal mechanics. Because of this, patients with either of these diseases cannot really get better without treating and subsequently eliminating these mechanical problems.

Later, I thought that perhaps by taking things a step farther and more fully optimizing structural alignment, sports performance could be increased. The results were equally gratifying.

I've been privileged to treat some of the greatest athletes in the world... Olympic gold medalists, world record holders and professional athletes and have found that *simple improvements in structural alignment can have a dramatic and often immediate effect on enhancing human performance.* These principles can be applied to everyone...even the elderly.

Over the years, I found that whether I was treating a child who was thought to have "growing pains," a teenager with runner's knee, young surgical patients with early osteoarthritis, a diabetic, or a world-class athlete... *finding and correcting underlying structural abnormalities had a profound effect on my ability to help them.*

I was taught about arthritis by some of the best minds in medicine. It was my initial area of interest and has remained so even as I have specialized in other aspects of medicine. I first published on arthritis in 1976 [6, 7, 8, 9] and in 1986 became one of the few doctors in the country in my specialty to become a Founding Fellow of the American College of Rheumatology. I have continued to treat arthritic patients while applying what I have learned from my studies in other fields of medicine.

TEN COMMON MISCONCEPTIONS

As I've mentioned, some of what we've learned about the causes and treatment of arthritis is currently being challenged. And as physicians, we must always question what we know so that advances can be made. Over more than 40 years of studying, observing, teaching and treating arthritic patients, I've consistently noted things that have made me question what I have learned about osteoarthritis. *This has led me to develop some principles and techniques that can help you...principles that are now being validated by studies conducted at some of the finest institutions in the world.*

But first, let's look at some of these misconceptions. For example, the foot is often not mentioned as a primary site of osteoarthritis, but it is indeed. Ask any podiatrist and you will learn that changes consistent with osteoarthritis are routinely seen on X-rays of the foot. Studies validate this (e.g., [10]).

Here are some of the most common and important misconceptions:

1. OSTEOARTHRITIS IS AN INFLAMMATION OF THE JOINTS

The word *arthritis* is credited to the Greek, Hippocrates, who named it combining forms *arthro-*, meaning "joint," and *-itis*, meaning "inflammation." Literally translated, osteoarthritis means "inflammation of the joints." As you have seen in the last chapter, *rheumatism*, is also derived from the Greek and means swelling. But unlike rheumatoid and other types of arthritis, there is usually little inflammation associated with osteoarthritis. Instead, joint pain and stiffness are the principal signs. *So, the very definition of our subject matter is really somewhat misleading.*

2. AGE IS THE PRIMARY CAUSE OF OSTEOARTHRITIS

As noted earlier, osteoarthritis is generally assumed to be a normal, unavoidable part of the aging process. Age itself is considered to be its primary cause. This seems logical. After all, even if you run water over a rock for a long enough period of time, it will wear away the rock's surface. So, if you live long enough, daily friction and pressure will simply wear out your joints as well.

Assuming that age is the primary cause of arthritis, is like assuming that age causes heart disease because many older people have heart problems. However, you don't get heart disease from the last potato chip you ate when you were 73. A logical error is at work here. A *correlation* is not the same as a *cause. In both heart disease and osteoarthritis, the pathological process actually begins in childhood, but it takes years before symptoms and tests can confirm the diagnosis.* By then a great deal of damage may already have occurred (See Figures 2A, and 2B).

In almost all instances of osteoarthritis, one knee or hip joint begins to hurt a patient first. Sometimes, that is the only one that ever shows symptoms. *Certainly, one knee is not younger than the other.* And in cases where both knees and hips are involved, the onset of one usually precedes the onset of the other. The reason for this is that pain and limited motion of the first arthritic joint causes an altered gait, and puts more stress on the other side, ultimately wearing it out also [11].

Why do some people have arthritis in their knees and not in their hips? Aren't those joints the same age? If age is the primary cause of osteoarthritis, why do some 90-year-olds have no arthritis in their knees, while other individuals much younger, have severe arthritic changes in these same joints? There must be another reason...and there is. *Abnormal alignment creates more friction on a joint and causes it to wear out faster (See figure 2C).*

Although there are certainly many factors involved, it is generally accepted that the average car's tires will wear out in about 40,000 miles. It can also be assumed that if you live to be 80 or 90 years old, some of your cartilage will also wear out. So age certainly is a factor. But you also know that you can drive a car just 10,000 miles with tires in poor alignment, and get much greater wear than you may have at 40,000 miles with properly aligned tires. That's exactly what happens to us (See Figures 3A, and B) ...we get osteoarthritis of a particular weight-bearing joint because *improper alignment, like a flattened foot or a longer leg, causes increased friction and pressure on a localized area of that joint.* You will come to understand the important, specific mechanisms of why this develops in later chapters.

We tend to think that osteoarthritis primarily occurs because of advanced age, as seen in Figure 2A. The arthritis in this patient's right knee actually began in childhood. Figure 2B shows the very same foot stance in this nine year old child, whose flattened right foot is causing a similar inward rotation of their right knee.

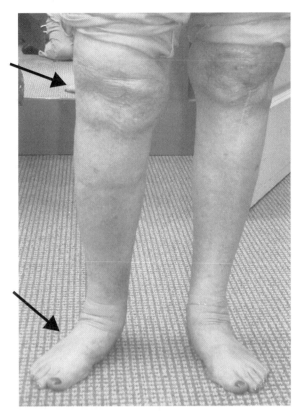

FIGURE 2A

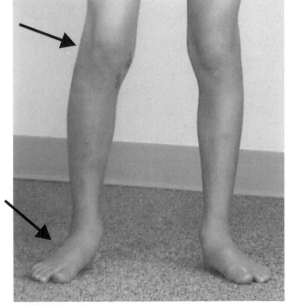

FIGURE 2B

Abnormal alignment can most always be seen on our shoes, just as it can on the tires of your car. The abnormal frictional forces that cause such wear patterns have a powerful deteriorating effect on our weight bearing joints.

This concept of abnormal alignment is critically important in preventing and eliminating the symptoms of osteoarthritis. As you will see, it is the cornerstone of this disease and the common link between its real cause and proper treatment.

3. EXCESS WEIGHT CAUSES OSTEOARTHRITIS

Second only to age, weight is consistently noted as a major cause of osteoarthritis. Certainly, the more weight you carry the more stress you place on the weight-bearing joints of your feet, ankle, knees, hips and back. The studies I cite in Chapter 3 validate this.

But here again…a logical fallacy is at work. Some significantly overweight people have no joint problems, whereas others, who weigh far less than average, have arthritic joints. Just as with age, many overweight people with arthritis have one knee or hip that hurts more than the other. Again, sometimes it's the only one that's ever involved. Is all their weight on one side? And why, if the overweight issue is resolved and an arthritic joint is replaced with a brand new one, does that joint usually need replacing at some point in the future?

So, like age, weight can *accentuate the problem…* but is not the *primary* cause. Additional weight can cause a joint to deteriorate because it increases friction and pressure. However, *friction and pressure that are concentrated in a localized area due to poor alignment, can cause a joint to deteriorate much faster and much more severely than additional weight that is evenly dispersed.*

If you've ever tried to hike with a heavy backpack, you can certainly appreciate this. Equally distributing the weight makes the pack a lot more comfortable and significantly less tiring to carry. That's why there's a specific way to pack it and books that teach us how to do this. Try putting a 30-pound dumbbell on one side of that backpack, and it won't take you long to feel the difference. Additional weight becomes much more important when the added load is placed on an uneven and improperly aligned structure, as in the case of a person with one leg longer than the other.

Research has validated this. In 2007 in an article in the *Journal of Bone and Joint Surgery*, researcher Bassam Mastri [12] reported finding a "higher risk of developing osteoarthritis in obese patients with poorly aligned knees than with normal alignment." Even a Rolls Royce doesn't ride well with a 20-inch wheel on one side and an 18-inch wheel on the other.

Abnormal weight also changes the way you walk and function. It is this change...rather than the extra weight itself...that is often the problem. Pregnant women, for example, walk quite differently to support their sudden extra weight and often develop joint problems because of it.

The reverse is also true. If you have significant structural problems, it's much harder to be active, causing you to gain weight and possibly develop arthritis. However, the abnormal structure, not the weight itself, is still the underlying cause of the arthritis.

4. FOOT POSITION HAS LITTLE TO DO WITH ARTHRITIS OF THE KNEES AND OTHER WEIGHT-BEARING JOINTS

To the contrary! *Although the foot really is not thought to be of much impor- tance in medicine, as I will show you in chapter 4, it is not only the foundation*

 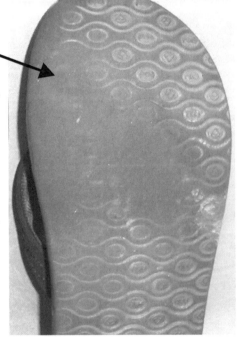

FIGURE 3A   **FIGURE 3B**

Abnormal alignment can most always be seen on our shoes, just as it can on the tires of your car. The abnormal frictional forces that cause such wear patterns have a powerful deteriorating effect on our weight bearing joints.

of our entire skeletal system, but when poorly aligned, it is the major cause of arthritis of the weight-bearing joints.

5. OSTEOARTHRITIS IS INHERITED

Because osteoarthritis seems to run in families, genetics is thought to play a role. Remember though…no "arthritic gene" has been identified for osteoarthritis. What you do inherit, however, is not just the color of your mother's hair or the shape of your father's nose, but the abnormal structure that caused the same problems they had. For example, if your father had arthritis of his right hip, you didn't inherit an "arthritic right hip gene." Instead, you may have inherited the same longer right leg or flattened foot that caused the increased friction and pressure on his hip.

Reviewing family pictures often confirms this. Generations can frequently be seen standing together with the same lower shoulder, bent knee, and flattened foot…similar structural abnormalities…that later caused their arthritic joints (See Figures 4A, B and C).

Bunions are another example. A bunion looks like an enlarged bump on the inside of the foot, but is actually a dislocation of the big toe joint. Often, because parents and their children have this same problem, bunions are thought to be inherited. But what is actually inherited is the tendency to pronate. This in turn dislocates the joint, subsequently causing the bunion (See Figures 5A and B).

I have consistently seen that by stopping the dislocating force of pronation, a bunion can be prevented. In such cases, I certainly haven't made an arthritic gene disappear, but instead, have removed the true underlying cause, which is the abnormal alignment (See Figures 5C and D). Medical data confirms my thinking. In an article in the *Annals of Internal Medicine,* reporting on a National Institutes of Health (NIH) conference held in 2000, conference chair David T. Felson [13] stated that, "even if someone has an inherited predisposition to develop the disease,…[it] will develop only where a biomechanical insult has occurred."

Figure 4A is a family portrait that includes a grandfather, whose right big toe joint was replaced by me 33 years ago due to severe osteoarthritis. The cause of his arthritic problem was excessive pronation or flattening of his foot combined with an old soccer injury. Upon careful inspection, his grandson can be seen to have the very same flattened foot that later contributed to his grandfather's problem (See Figure 4B). Figure 4C shows a mother and daughter who have the same flattened feet and arthritic knees due to the limb length discrepancy they each have.

FIGURE 4A

FIGURE 4B

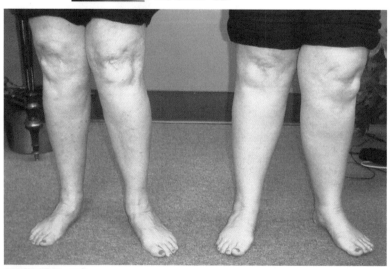

FIGURE 4C

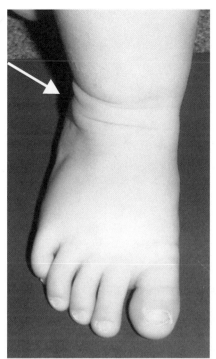

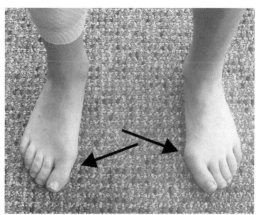

FIGURE 5B

Long before an actual dislocation of the big toe joint occurs, pronation can be seen. Figure 5A shows pronation in a toddler. Figure 5B shows the progression from untreated pronation to a severe bunion in the left foot of an eleven year old. The difference in the severity of the deformity seen in this child's right and left feet is directly related to the degree of pronation.

FIGURE 5A

6. IF YOUR JOINTS DON'T HURT, YOU DON'T HAVE ARTHRITIS

While joint pain is usually an indication that you do have a problem, the converse is not necessarily true. Just as you can have significant heart disease and not have pain, you can have joint damage and initially remain symptom free. So the absence of pain may simply mean that the arthritis is still in its early stage and not severe enough to cause discomfort.

Sometimes stiffness or a limited range of motion that you don't even realize you have, can be an indication of arthritis. Remember too, that if you are taking pain medications, anti-inflammatory drugs or other types of remedies, your arthritis may be progressing despite symptomatic relief. As we saw in the last chapter, certain medications can actually be making your arthritis worse in spite of the fact that your symptoms may seem to be improving.

7. USING A JOINT KEEPS IT HEALTHY

The cliché, "Use it or lose it," is as true here as it is with most other parts of our bodies. Among other things, using a joint increases the thickness of hyaluronic acid in our joint fluids, which is necessary for good joint lubrication. Exercise also keeps the muscles around our joints strong and healthy, which is essential for joint stability. Further, when inflammation occurs in a joint, small proteins are produced. Exercise counteracts this process because the increased blood flow helps remove them, and bring healthy nutrients to the joint, thus reducing pain and stiffness.

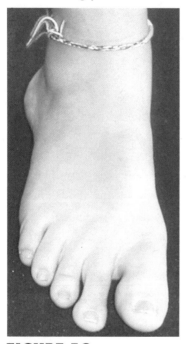
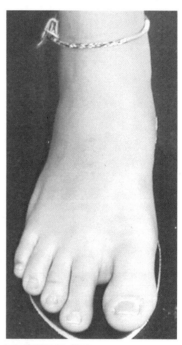

FIGURE 5C   **FIGURE 5D**

This child's severe pronation causes her heel to become laterally dislocated (See Figure 5C). When the foot is held in a corrected position this prominent dislocation is not evident (See Figure 5D). In both figures, notice the space between her big and second toes. The big toe is in a much better position and further away from the second toe in Figure 5D, when her pronation is controlled by the use of a foot orthotic. The rotation occurring at the big toe also changes with correction. Notice the straightened position of this patient's big toenail in the same picture.

So, moving a joint is good...but only if you are doing so in proper alignment! Moving a joint in poor alignment will cause further pain and joint deterioration. This is especially true if the motion is very repetitive and performed under great stress. We need only look at a pitcher's elbow to be reminded of this.

I often see people exercising not only to keep their joints healthy, but for other medical benefits...such as losing weight, and lowering their cholesterol and blood pressure. Often times, they are unknowingly exercising in poor alignment and are doing far greater damage than good (See Figure 6).

Don't get me wrong...I'm a big advocate of exercise and certainly practice what I preach in this regard. But ideally, weight-bearing exercises should be halted and only resumed after joints are properly aligned. Then you may be able to do much more than you ever have before.

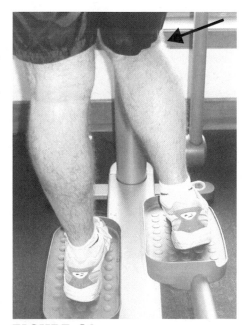
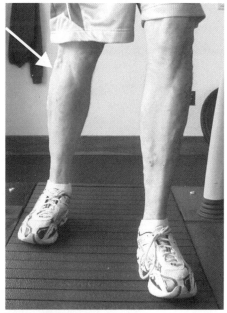

FIGURE 6A   **FIGURE 6B**

> While these patients may be lowering their weight, cholesterol and blood pressure by exercising, doing so with a knee turned inward, (as these patients' right knees are), will cause excessive pressure on that joint and deteriorate it. This is true with any piece of exercise equipment if the patient functions in poor alignment.

8. REPLACING A KNEE OR HIP JOINT FIXES THE PROBLEM

Replacing an arthritic joint doesn't fix the problem anymore than putting a new tire on a car with a bent frame fixes the frame, gives you a better ride, or helps the tire last any longer. As a matter of fact, as I've mentioned, *one of the major complications of knee, and especially hip joint replacement surgery, is a limb-length discrepancy...that is...ending up with a leg that is even longer or shorter than before the surgery;* sometimes that difference can be quite significant.

That's because typically, whenever a joint is replaced, the length of one or more of the bones involved in that joint will also change. As a former surgeon who considers himself something of a perfectionist, and has done many joint replacements in the foot over the years, I can tell you this is certainly true. So, regardless of what you're told...making your legs exactly even isn't possible with hip replacements. If the good Lord didn't make your leg lengths perfectly even, no surgeon will either. Despite the critical important implications of this fact...equalizing leg lengths is usually ignored or poorly followed up.

This not only causes stress on the replacement joint, often necessitating its replacement within a few years, but it wreaks havoc on your other joints. Indeed, research has validated that failing to properly align a patient after surgery results in a less successful surgical outcome [13].

Even in the best of circumstances, joint replacement surgery is not a definitive answer to osteoarthritis.

9. SUPPLEMENTS LIKE GLUCOSAMINE AND CHONDROITIN SULFATE CURE ARTHRITIS

These popular supplements may help replace damaged cartilage, which can result in decreased symptoms, but as mentioned, they do not prevent or eliminate the cause of the arthritis. Nor do they prevent or treat its associated joint problems, such as tendonitis or the muscular splinting and guarding that often accompany the disease. Since abnormal alignment is the primary cause of arthritis, if and when supplements reduce symptoms, a false sense of well being may occur, because joint erosion may be developing without your knowledge.

10. ORTHOTICS HAVE LITTLE EFFECT ON ARTHRITIS

As you will see in Chapter 7, a great many misconceptions exist regarding the use of foot inserts in arthritis. For now, and regardless of the fact that you may have had one or more bad experiences...suffice it to say...that they can indeed have a very positive and powerful effect on both preventing and decreasing arthritic symptoms. A significant amount of credible data which I will present supports this premise.

THE BOTTOM LINE

I'm certainly not trying to prove that "they"... the medical establishment... are all wrong, and I'm all right. If I had all the answers, I would be writing to you from my own private island instead of my office outside of Atlanta.

Here's the bottom line...age and excessive weight are indeed important factors...but they are *secondary* ones. If and when you develop osteoarthritis depends on a combination of factors related to abnormal alignment. Someone with ideal alignment may be overweight and not develop

symptoms, yet someone with major structural problems who is very active and of normal weight, may develop arthritis at a very young age. And of course there are always exceptions, like the three pack-a-day smoker who never develops problems.

> "*A complication of knee and hip replacements is most often a leg that is longer or shorter than it was before surgery. Equalizing leg lengths creates balance and stability and decreases further joint damage. I'm an orthopedic surgeon who has had his own knee replaced. But as a patient of Dr. Pack's, I've been most impressed with his unique techniques to correct this and other structural problems. They are invaluable for those like me with arthritis, as well as any athlete who wants to maximize their performance.*"
>
> Butch Fossier, MD
>
> Board Certified Orthopedic Surgeon; Former Team Physician, Chicago Bears, 1978–1991

CHAPTER 3

The Real Cause of Arthritis

PREMISE and VALIDATION

"I have done a great many total knee replacement surgeries for arthritis. Dr. Pack has helped me more fully appreciate the important role alignment plays in causing those arthritic joints. My own left knee has plagued me for years when doing surgery, causing knee and back pain. I attributed it to age, an old ACL injury, and the stress of my work. Thanks to him, I can now operate all day with minimal pain in my knee or back. I believe his book, "the Arthritis Revolution," will be an invaluable source of information for all of those afflicted with arthritis, as well as for us, their caretakers."

Tim Stapleton, MD

Board Certified Orthopedic Surgeon specializing in joint replacement surgery, Clinical Assistant Professor, and Head Team Physician, Mercer University, Chief of Surgery and the Chief of Orthopedics for Coliseum Northside Hosptial. Orthopedic Team Physician, 1995 Atlanta Cup, and 1996 Olympics

- Pack's Postulates explain the role of abnormal mechanics in arthritis of the weight-bearing joints.
- Structural abnormalities cause abnormal function which results in increased joint stress.
- Studies now validate the premise that the increased joint stress resulting from structural abnormalities causes arthritis of the weight-bearing joints

PACK'S POSTULATES

The role of abnormal mechanics in arthritis and sports performance is based upon three basic principles:

1. No one is born structurally perfect; i.e., we all have structural abnormalities. Even our right and left sides are not absolutely identical to each other. Simply look in the mirror, and you will see that one shoulder is higher, one arm lower or one foot flatter than the other.

2. Any degree of abnormality... like a flattened foot, or longer leg...whether congenital or acquired from an injury or a primary arthritis like rheumatoid arthritis, causes improper alignment which:
   A. Increases joint friction and the risk of injury, which can lead to osteoarthritis
   B. Decreases optimal function, including sports performance

3. Conversely, correcting these abnormalities can, and often does, have a profound and sometimes immediate positive effect in reducing symptoms by decreasing joint stress, and the risk of injury, as well as increasing sports performance. Much of this can be accomplished by optimally aligning the foot...which is the foundation of our entire skeletal framework.

HOW STRUCTURAL ABNORMALITIES CAUSE ARTHRITIS

As the following diagram shows, structural abnormalities whether congenital or acquired, can result in malalignment. This is the primary cause of osteoarthritis of the weight-bearing joints, which as mentioned, are the feet, ankles, knees, hips, back, and neck. In people who participate in sports, decreased performance and increased risk of injury precedes arthritic changes. With additional weight, age, or activity, arthritic changes generally become more severe and can occur earlier.

The result of the sequence shown in Figure 7 can be seen in the patient in Figures 8A and B. This woman has both congenital and acquired structural abnormalities. She was born with a longer right leg which became even longer as a complication of her total right knee and hip joint replacements. The result was

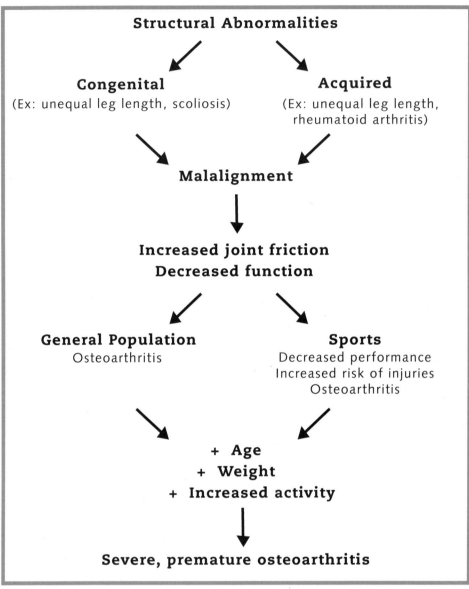

FIGURE 7

significant malalignment, which in turn, increased friction in these joints and decreased function. This subsequently caused osteoarthritis not only in her right knee, which was initially affected, but in all of her other weight-bearing joints. Statistics show that she will most likely need another replacement of her artificial right knee. The longer she lives, the more weight she carries, and the greater her activity level, the more severe her arthritis will likely become.

RESEARCH VS. CLINICAL EXPERIENCE

Over the years, I have developed a method of evaluating and treating patients that has proven to be effective in treating thousands of cases of arthritis, as well as in increasing the performance of some of today's greatest athletes. But quite frankly, as valid as my work is clinically, it is still regarded as anecdotal data by some in the medical community, and sometimes doesn't hold much credibility with the research folks. For them to consider results valid, premises and theories must first be proven in carefully conducted clinical trials, in unbiased settings and according to very strict guidelines. Once validated, a report on that work can be compiled and submitted for publication in a credible medical journal. If and when it is accepted and other promi-

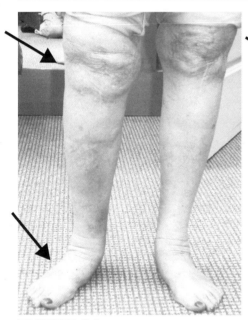 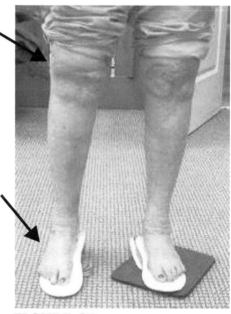

FIGURE 8A

BEFORE TREATMENT
Notice her longer right leg. Her right knee is bent and turned inward as she unknowingly attempts to shorten it. Both feet are significantly flattened and rolled inward.

FIGURE 8B

EVALUATION
Simply standing on a lift under her shortened left leg equalizes her leg-length discrepancy. The addition of optimally angulated foot inserts corrects her pronation and provides structural alignment.

nent, independent researchers conduct their own studies and substantiate the premise, it may eventually be incorporated into routine medical practice. This process, although often a necessary and valuable safeguard for medical judgment and practice, can take years...often outliving the physicians who originated the premise and in the meantime, deprive many of much needed care. At other times, regardless of how important a discovery is or how effectively it *has been proven*, medical acceptance *can still be denied* for long periods of time. For example, guidelines for the practice of medicine were originated and published in the medical texts of ancient Greece for 2,000 years, yet resistant thinking prevented this same material from being disseminated during that entire period [1].

Perhaps the most important medical breakthrough in modern times came from Dr. Ignaz Semmelweis, who noticed that there was a high incidence of death in mothers from childbirth fever, among the patients of doctors who had just performed autopsies and then delivered babies, without first washing their hands. Despite the fact that he was able to prove a significant reduction in such deaths after instituting hand washing, Semmelweis' ideas were so ridiculed, that he was eventually committed to an asylum and died penniless. It wasn't until some 80 years later, in the 1920s, that sterile gloves became widely used.

Semmelweis' legacy encompasses far more than his incredibly significant medical breakthrough, which is now immortalized with the term *Semmelweis reflex. This term refers to the automatic rejection of new medical information, regardless of how sound or proven, simply because of narrow-minded, deep-seated ideas [2].*

Clinicians, as opposed to research scientists, are less interested in what works in a laboratory or with test cases, than they are in the effectiveness of treatment routinely used on their own patients. In the real world of medicine, the fact is that many physicians simply use methods and techniques because they know they work, not because scientific testing has demonstrated their validity. This is exactly why, although they depend on both in vivo (inside the human body) and in vitro (in an artificial environment) studies, physicians generally have more faith in the former... that is, clinicians are generally more interested in how things work in the real world.

As a matter of fact, many things done in medicine have never been scientifically proven, but are known to be effective and are therefore accepted. In other instances, products and drugs are found useful for purposes other than those for which they were originally designed. For example, the anticonvulsant drug topiramate (Topamax), which was primarily intended to relieve convulsions and migraine headaches, is now used to help alcoholics stay sober.

Physicians also know that studies can be misleading, such as those that are funded by the companies that manufacture the drugs. In such scenarios, criteria and testing methods can be altered to show results favorable to the product and ultimately to sales. That is why it is so important to have independent studies conducted under stringent guidelines at reputable institutions. The reputations of prestigious medical journals such as the *New England Journal of Medicine* and the *Journal of the American Medical Association* are dependent on the publication of studies conducted in this manner.

Ultimately, both research and clinical data are necessary, as is open-mindedness. So, before we go any further, despite the fact that (as I have mentioned), my methods have been tested in thousands of cases over forty years, since they are not yet standard practice in the arthritis community, I think that it is important for me to present the hard data that supports the premise of my work… that *abnormal alignment is the major causative factor in arthritis of the weight-bearing joints.*

When I began looking for such corroborative evidence for this book, I was quite pleased to find extensive, supportive research that had been conducted at some of the finest institutions and published in our most prestigious and highly respected medical journals. Some of this data I have already referenced. Other pertinent sources will be cited in later chapters. Much of what follows is related to arthritis of the knee which is likely to affect *one out of every two senior citizens.* Please know that the premise of these works can readily be applied to the other weight-bearing joints as well.

It should be noted that this is but a mere sampling of the supportive data available. A comprehensive review would require quite a lengthy manuscript of its own. Furthermore, most of the sources I cite have relied on extensive, supportive references, which they too, have documented.

VALIDATION

The latest research clearly and increasingly emphasizes the important role of pathomechanics (abnormal structure and function) as a causative factor in the development and progression of osteoarthritis of the weight-bearing joints.

In the 1970s, researchers hypothesized that increased, repetitive joint-loading (such as that due to abnormal alignment) could cause osteoarthritis [3, 4]. A series of animal studies later confirmed this [5, 6, 7]. According to Gross and Hillstrom [8], "nearly all patients who have symptomatic knee osteoarthritis, report that their pain is provoked by some kind of weight-bearing activity" (p.756).

Because the knee is the most common joint involved in osteoarthritis, much research continues to be conducted on this joint, some of which is summarized in the following paragraphs. As mentioned earlier, an increasing number of studies are finding that structural malalignment is a prime factor in both the cause and progression of knee osteoarthritis [9, 10, 11].

It has also been shown that lower limb malalignment redistributes loads on the knee joint [12] and that rotational forces can cause asymmetric pressure, resulting in cartilage damage in weight-bearing joints [13]. In 2003, researchers reported in the *British Journal of Sports Medicine* [14] that "biomechanical factors are likely to contribute to the causes of osteoarthritis of the knee," and further stated that "these causes are becoming increasingly recognized."

Tanamas and colleagues, in systematically reviewing the literature, concluded that knee malalignment altered the pressure across the joint and increased cartilage damage and was therefore, a factor in causing osteoarthritis of the knee. They further stated that there is "strong evidence" that this malalignment is a risk factor for the progression of osteoarthritis that is seen on X-rays [9].

Abnormal knee alignment has been associated with the risk of joint space narrowing and the formation of osteophytes (bony outgrowths; [15]) as well as the rate of cartilage loss in those with knee osteoarthritis......all characteristics of this disease [16].

Much work has been done to show the effects of structural abnormalities such as tibia varum (bowleg) and tibia valgum (knock-knee) on arthritis of the knees. Abnormal alignment in bow-legged individuals increases the stress on the medial (inside) compartment of their knees, and this is thought to contribute to the wear seen in their articular (joint) cartilage [17, 13].

In 2001, Sharma and colleagues [17] published the results of a study that they believed showed for the first time that *abnormal alignment of only 5 degrees (as measured from the ankle to the hip) increased the progression of osteoarthritis 3 to 4 times,* as compared with neutral or good alignment. This could be seen as soon as 18 months after being identified. It was also found that the greater the malalignment, the faster and worse the progression. *The researchers concluded that osteoarthritis was, therefore, a result of local, mechanical factors.*

In 2008, researchers at the Mayo Clinic Motion Analysis Laboratory confirmed the results of previous studies, which indicated that the risk and rate of progression of knee osteoarthritis were increased by leg malalignment. *Their own study showed that for each degree of valgus malalignment (in people who already had some arthritis), there was a 55% increase in the risk that the outside compartment of the knee was affected by osteoarthritis. They also found that "increasing age was only weakly associated with an increase risk of osteoarthritis of the knee."*

This is quite important, as it confirms my opinions about the relationship of age to osteoarthritis. They concluded by saying that *"alignment has been cited as one of the most important risk factors for the progression of osteoarthritis"* and that *"there is a clear relationship between overall limb alignment and arthritis of the knee"* [18].

Defects in joint cartilage can be an early sign of osteoarthritis and the probable need for total knee replacement. An article in the *Journal of Orthopedic Research* stated that malalignment increased joint space narrowing [16]. And in a study published in the journal *Rheumatology,* Cicuttini et al. [19] reported increased cartilage damage in those with valgus knee alignment. A number of studies have also confirmed that knock-kneed patients have a higher risk of osteoarthritis [17, 18, 19].

Dr. Jason Theodosakis is considered an authority on osteoarthritis including biomechanical interventions. More than one million copies have been sold of the first edition of *The Arthritis Cure,* [20] which he coauthored in 1997, with Brenda Adderly and Barry Fox. Although the book's major focus is on the supplements glucosamine and chondroitin sulfate, throughout this popular book, which is based on more than 150 documented research studies, the authors continually emphasize the important role that biomechanics and abnormal function play in the development and treatment of osteoarthritis. In a chapter entitled *The Arthritis Cure,* the authors state, *"The importance of biomechanics (the study of the mechanical forces exerted on the body by movement) in treating osteoarthritis can't be overstated: If you don't correct the underlying problems, you can't rid yourself of the disease" (p. 84).* These authors also agree that because poor bone alignment causes osteoarthritis, the same principles that can decrease the progression of osteoarthritis once it's begun, can also be used to prevent it [20].

Dr. David Hunter [21], a preeminent rheumatologist and director of research at New England Baptist Hospital and Boston University School of Medicine, was asked by the medical quarterly, *Rheumatic Disease Clinics of North America,* to review the published data, which consisted of an enormous number of studies for their August 2008 issue. Devoted exclusively to osteoarthritis, this issue provided a compilation of the best and latest research on this subject for health care providers.

In the preface to that issue, Hunter commented that *"Osteoarthritis is no longer viewed as a passive, degenerative disorder but rather an active disease process driven primarily by mechanical factors"* (p. xiii). He now believes that *"Mechanics play a critical role in the initiation, progression, and successful treatment of osteoarthritis"* (p. xiv). Dr. Hunter recommends that "We learn from the insights our research is providing to focus even more on important modifiable risk factors, such as mechanics..." and adds "that by so doing, we have the opportunity to make a difference in millions of peoples' lives" (p. xvi).

Reporting in the same journal, other researchers fully concurred. David R. Wilson and his colleagues stated, *"There is little doubt that mechanics play a role in the initiation, progression and successful treatment of osteoarthritis...*

much like bearing surfaces in an engine that have worn down after too many revolutions" [22, p. 605]. After an extensive review of the medical literature to 1999, they found significant evidence of the role of abnormal alignment in osteoarthritis and stated that, *"Since that time, the importance of joint alignment has become even clearer"* (p. 612).

In their article "Epidemiology of Osteoarthritis," Yuqing Zhang and Joanne M. Jordan wrote, "Knee alignment (i.e., the hip-knee-ankle angle) is a key determinant of load distribution" and speculated that any abnormality of this angle would increase pressure on the knees and subsequently result in both a higher risk of developing arthritis and the progression of this disease [23, p. 523]. In reviewing the medical literature, Zhang and Jordan found a number of studies confirming their hypothesis and referencing the important relationship of abnormal alignment and osteoarthritis.

In 2009, the results of a systematic review of the literature, which concluded that malalignment of the knee joint was an independent risk factor for the progression of osteoarthritis, [9] was published in *Arthritis and Rheumatism,* the official journal of the American College of Rheumatology. *One comprehensive study tested more than 1,500 participants who were followed for more than 6 years and found a positive association between poor alignment and increased development of osteoarthritis.*

According to their website, the Arthritis Foundation is, "the largest private, not-for-profit contributor of arthritis research in the world." It provides public education on arthritis and works with other credible organizations, such as the Centers for Disease Control and Prevention (CDC). In April 2009, the CDC and the Arthritis Foundation organized expert working groups to develop white papers for discussion at an osteoarthritis summit meeting. More than 80 people from 50 organizations came together to discuss the status of public health interventions for this disease. *Their report confirmed the undeniable importance of mechanical factors in relation to the symptoms and progression of osteoarthritis* [24] (See Figures 9A and B).

With reputable, plentiful data such as this now available, there is no question that abnormal alignment is an important contributing factor in the cause and progression of osteoarthritis. Patients should be evaluated to assess stance and function. Then, every effort should be made to correct any abnormalities in order to decrease joint stress and optimize function. This would enable physicians to treat osteoarthritis at its root cause and not simply medicate it after the fact. No longer should such evaluations be omitted from any examination by a physician, and others entrusted with treating a patient's arthritis. Doing so renders even the best therapy incomplete.

The concept that age is the primary cause of arthritis has to change for you to get the care you now need. As mentioned earlier, the change should begin

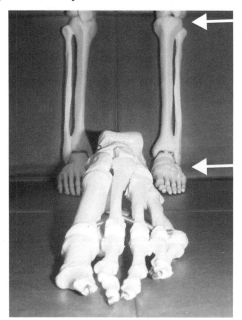 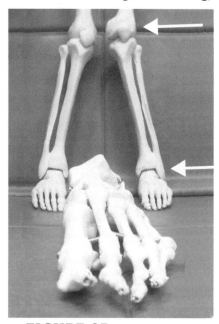

FIGURE 9A   **FIGURE 9B**

Figures 9A and B are examples of how foot position affects the alignment of the weight bearing joints above it. Figure 9A shows that with proper foot positioning, the knees can also be aligned, whereas in Figure 9B, when a foot is pronated, the knees turn inward causing a sequence of improper alignment all the way up the skeletal chain. If you refer back to Figures 1C and D, you can begin to appreciate how abnormal alignment can cause increased friction and pressure on a joint, thus causing its deterioration.

with pediatricians who are our first line of defense in preventing flexible, often easily corrected deformities from becoming fixed, rigid ones. The change should also include primary care and family practice physicians, internists, orthopedic surgeons among others. Most of all, it is my fervent hope that more rheumatologists will now embrace this important, proven material and begin to include structural analysis in their evaluations of arthritic patients.

"*Orthopedic surgeons told me I had 'bone on bone' arthritis due to years of basketball injuries, and that I needed knee replacements. As a coach, I couldn't even warm up with my players without severe, debilitating pain. Now, after being treated by Dr. Pack, I can teach and even play without any pain. So I had him treat my whole team so they wouldn't end up like I did. Those that were having significant problems are all well now and I can definitely see the increased performance they each have. Seeing the results, the head team physician, (an orthopedic surgeon), who had knee pain himself, went to Dr. Pack, and can now stand and operate with minimal discomfort. Anyone who wants to decrease joint pain and or improve sports performance must see this guy!*"

Janell Jones,
Women's Head Basketball Coach, Cal State

Former Head Coach, Mercer University, University of California, San Diego and Oklahoma City University (the winningest coach in OCU history), two-time national Coach of the Year (winning over 90% of her games in seven seasons), two NAIA national championships, California Collegiate Athletic Association Coach of the Year.

CHAPTER 4

Our Structure

WHERE the RUBBER MEETS the ROAD

"You simply cannot lift heavy weight or maximize your strength without a good foot stance. It is not just important but critical. Dr. Pack is the master!"

Robert Stephens

Certified Personal Trainer and Conditioning Specialist, former Personal Training Manager, Lee Haney's Enterprises; 3-time Natural Powerlifting Champion from 1995-1997

- Ideal alignment begins with a stable foundation.

- The feet are the foundation of the body and all of its weight-bearing joints.

- The subtalar joint, which lies below the ankle, is the structural core of the human body.

Now that you have seen that abnormal alignment...and not age... is the primary cause of osteoarthritis, let's look at where that alignment comes from.

OUR STRUCTURE

We are miraculous, complex entities. In an average lifetime the human heart will beat more than 2 billion times [1]. The lungs will inhale and exhale more than 600 million times [2]. With thousands of miles of blood vessels and nerves and a computer between our ears that remembers, calculates and actually thinks and reasons for itself, we are indeed remarkable.

We are also structural beings with our organs given form by a framework of 206 bones. But like all things in this world, we are not made perfectly to begin with, and are then subjected to the deteriorating effects of friction and pressure. Like a rocky coastline repeatedly washed by soft waves, our joints will wear away to some degree over time.

As previously stated, poor alignment is the primary cause of the increased friction that deteriorates our joints. Indeed, wherever there are moving parts, alignment is critical to sustained optimal function. That's why if the builder of a car engine wants the hardened metal cylinders ...regardless of what they are made of...to last longer and perform better, the engine will be balanced and blueprinted to exacting specifications within thousandth-of-an-inch tolerances.

Also, as I have mentioned, it is the same reason manufacturers stress the importance of periodically aligning your car's wheels...to lessen abnormal friction on the tires. Alignment results in longer tire life and a better ride. *The same is true of your joints. The better they are aligned, the longer they will last and the better the ride...the less discomfort...you will have throughout your life.*

Think of your body as a building with many rooms. Your heart, lungs, stomach, and so forth are the rooms in that building. Like the concrete edifice with its steel reinforcements that supports the building and those rooms, you have an intricate stabilizing system giving you much needed support... your skeletal system. But unlike a building... *you move.* Supported by ligaments, your 300 joints enable your structure to transport itself at a second's notice, with speed and balance over uneven surfaces, by means of a complicated system of hundreds of muscles and tendons. With a system as sophisticated as this, is it any wonder that we are not perfect?

---

[2] This number represents an average. The particular number depends upon how one technically classifies a joint because some move while others don't.

A building's supportive framework begins with and rests on its foundation… and so does yours…your feet. *If your feet are improperly aligned, all the weight-bearing joints they support will also be out of alignment.* The Leaning Tower of Pisa didn't get that way because of a solid foundation!

Realizing the importance of a building's foundation, a lot of time, effort, and money are spent to make it as strong and as stable as possible. Specialists… structural engineers…are employed to make sure the foundation is level and solid. And if you have ever watched a foundation being built for a large building, you may have noted the pains that were taken to make sure the foundation was as close to perfect as the builders could possibly make it.

But we cannot control how *our* foundations are built. Good or bad, rich or poor, we simply arrive without having had a say in the matter. However, unlike a building whose foundation is difficult, if not impossible, to change once it is finished…ours can often be corrected quickly and precisely, thereby improving the alignment and decreasing the stress on all of the joints our foundation supports.

So, if we want to decrease the incidence of arthritis or reduce its symptoms, it is imperative that we optimize our foundation, and that begins first with our feet. Later, I will discuss how this is done and the profound effect it has on decreasing arthritic symptoms.

THE HUMAN FOOT
OUR ORPHAN APPENDAGE

The proper positioning and alignment of our feet helps us walk upright, separating us from other species. For humans, the feet are the foundation of the entire body and all its weight-bearing joints. The feet are where our "rubber meets the road."

These two appendages, often disdained as the ugliest parts of our bodies, are the entire basis of our vertical structure. If not properly positioned, no weight-bearing joints above them can be aligned either. Because of this, by properly aligning our feet we can improve many structural problems…especially arthritic problems of the joints the feet support.

Athletes know all too well the importance of foot positioning. Watch sprinters before a race and notice how gingerly and precisely they place their feet on the track and how many times they set and reset them. Baseball pitchers, discus throwers and weight lifters all painstakingly set their feet. A swimming meet can be won or lost by how quickly the swimmers' feet propel them into the water on the initial jump. All of these athletes' performances are first dependent upon the optimal alignment of their feet. Watch golfers and note how important they think foot stance is before driving a ball. They know the importance of balance and that it all begins and ends with the stability of their foundation.

Yet despite their importance, we probably pay less attention to our feet than to any other part of the body. The truth is most of us take our feet for granted, assuming they will be affected by little more than corns, calluses, thick toenails, and the occasional wart. We are apt to spend more time choosing which gel toothpaste to buy than thinking about our feet. Except for the pretty pedicures some women get, the dirty shower water that runs off the rest of our bodies is about all the care most feet ever receive…pitiful indeed. But just let them hurt and ground us, cause us to miss our Sunday afternoon golf game, or keep a professional athlete out of play, and oh boy…are we ever surprised. And a new appreciation for these two structures quickly develops!

If we really understood just how much we ask of our feet and how little attention they get, we wouldn't wonder why 87 percent of Americans have foot problems at some point in their lives [3].

FOOT FACTS

So let me tell you a little about these two "orphan appendages" we depend on every day. They are actually architectural marvels encompassing one quarter of all the bones in the body. There are 33 joints…an arthritis paradise!.. and seemingly endless ligaments, tendons, and muscles. The bottoms of these masterpieces have three layers of intricate muscles. So trying to remove a foreign body, like a piece of glass for example, from the bottom of the foot's complicated structure, is truly like finding a needle in a haystack. Some of the nerves and blood vessels are so tiny as to be invisible to the naked eye, and the mechanics involved in their function can take a lifetime to understand.

Our feet will carry us 10,000 to 15,000 steps per day or about 115,000 miles in one's average lifetime. That is the equivalent of walking four times around the earth [4]. And talk about pressure...with every step we exert half again as much as our body weight. This is equivalent to several hundred tons of combined weight over an average day.

When we run, our feet hit the ground with 3 to 4 times the force of our entire body weight. That is as much as 20,000 pounds of pressure per square inch or 127,000 pounds of pressure on our feet each mile [5]. By the time our life is over, we will have put more than 900,000 billion pounds of pressure on them! Just imagine driving that distance with poorly aligned tires on your car. So, if you hear a little voice from below saying, "How about a little respect?" it will not be Rodney Dangerfield...but your feet... talking to you!

THE FUNCTIONS OF THE FOOT

The foot has three major functions: when standing...to balance and support our vertical structure, and during our gait (walking) cycle...to adapt to uneven surfaces and then act as a rigid lever propelling us to our next step. Although many of the joints in the foot play an active role in gait, much of our mobility centers around the ankle joint, which allows our feet to move up and down, and the subtalar joint which lies below the ankle. It is the subtalar joint that allows the foot to be a mobile adapter through the rolling in motion of pronation, and a rigid lever through the rolling out motion of supination.

Although our discussion here is limited to the structural and functional aspects of the foot, it should be noted that our feet also play another very important and very much overlooked role as an indicator of disease. You may be surprised to know that many of the signs of various diseases occur in the feet first, and are best diagnosed by thoroughly evaluating them. In the words of Dr. Marvin Steinberg, my beloved mentor, "The foot, much like the human eye, is truly a mirror of systemic diseases." [6]

Diabetics, for example, often develop neurological changes in the feet consisting of burning or tingling sensations. Sometimes these symptoms precede the onset of clinical diabetes by many years. The foot can also be an indicator of vascular insufficiency or decreased blood flow. Pain in the arch can occur because of a condition called *intermittent claudication,* in which poor blood flow causes pain with exercise, that is similar to the pain of angina that often occurs in those with a heart condition. That's one of the reasons it is just as important…sometimes more important…to check the pulses in your feet as it is to do so in your wrist. Stress fractures in the foot may be the first sign of osteoporosis in postmenopausal women, while abnormalities of the toenails can not only develop with common fungal infections, but with heart conditions like subacute bacterial endocarditis, thyroid abnormalities, anemias, and other disease states.

Besides osteoarthritis, various other arthritic conditions also affect the foot. Heel pain occurs in rheumatologic disorders such as ankylosing spondylitis (an arthritic condition of the spine), and psoriatic arthritis (a combination of psoriasis and arthritis). The big toe joint is affected in the majority of cases of gout. Swelling of the lesser toe joints may indicate inflammatory types of arthritis like Reiter's syndrome, while sweating of the feet can be an early sign of rheumatoid arthritis.

Physicians should always be cognizant of the myriad of signs and symptoms that can affect the foot and indicate an internal disorder. As far-fetched as it may seem, I have diagnosed cases of cancer and brain tumors through foot evaluations.

THE SUBTALAR JOINT
THE STRUCTURAL CORE OF THE HUMAN BODY

As stated before, although still not generally accepted, more and more physicians are beginning to agree that abnormal alignment plays a significant role in osteoarthritis. The dilemma is that they are not sure just what to do about it.

Within the complex foundational structure of the human foot, a single, small joint bears most of the responsibility for the functions of stability and mobility. This critically important joint that most have never even heard of is the subtalar joint. Located just below the ankle, it consists of the long heel bone of the foot, called the calcaneus, and the talus, which is the bone found just above it (See Figure 10). *It is the optimal alignment of this critically important joint that is the answer to the dilemma.*

If you are interested in trying to locate your subtalar joint, first find the bump on the outside of your ankle bone. This is the fibula. Move your finger to its underside, and then forward about one and a half inches toward the front of your foot and ankle. You should now be on the lateral side of the subtalar joint. It will feel like a small, depressed, concave area that will open and close as you roll your foot inward and outward (See Figure 11).

Because of their close anatomic proximity, and the fact that the talus is a part of each joint, the subtalar and ankle joints are sometimes confused with each other… although they function differently. The ankle joint consists of two leg bones, the tibia or inside bone, and the fibula or outside bone, and the talus, which is the top bone of the foot. The ankle is a hinge joint which just allows you to move your foot up and down. The subtalar joint on the other hand, only moves with a rolling motion, along an arc, creating the inward (supination) and outward (pronation) motions that were mentioned above (See Figures 12A, B and C).

A MOVING WALL

If you were going to build a high brick wall, you would first need a very solid foundation. Then it would be critically important to make sure that each brick

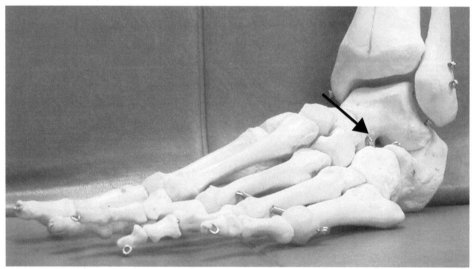

FIGURE 10

The subtalar joint lies just below and in front of the ankle as indicated by the arrow in this picture.

was optimally aligned or level. *This is especially true of the very first bricks...* *the bottom two. Because if the bottom ones are off center just a little...each* *subsequent brick on the wall will be even farther off.* Anyone who has ever built blocks with their children knows this all too well.

The bottom two bricks of our foundation are the talus and calcaneus and are *primarily responsible for stabilizing our entire skeletal system. They ultimately* *determine the static and functional positions of all the other weight-bearing* *joints of our bodies.* As we have seen, if not optimally aligned, all of the other joints from our ankles and knees up through our hips, back, and neck will be affected (See Figures 13A and B).

Our foundation not only has to provide maximum stability when we are still, but also must maintain that function as we move. When our feet need to be a rigid lever... as when we are lifting a heavy weight...the bones of the talus and calcaneus should sit perfectly positioned...one on top of the other... almost as if they were locked in place. This ultimate joint alignment is called *subtalar neutral,* and means that the joint is neither rolled in nor rolled out

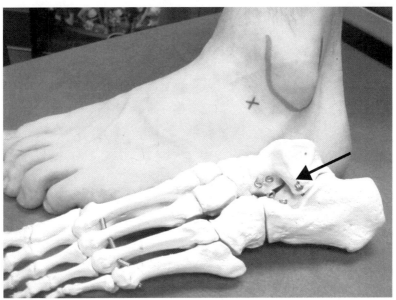

FIGURE 11

The X on this patient's foot and the arrow on the skeleton show the subtalar joint. The marked outline on the patient's foot is the fibula.

(See Figure 12A). At other times, when the foot needs to be a mobile adapter… such as when we are walking in sand…a certain degree of pronation and supination is necessary. But his motion must never be excessive, or it will wreak havoc on our skeletal system (See Figures 14A and B).

No other joint in our body so profoundly impacts so many other weight bearing joints as the subtalar joint. For this reason I believe the subtalar joint is the "structural core of the human body."

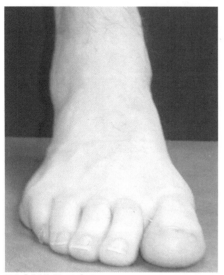

FIGURE 12A

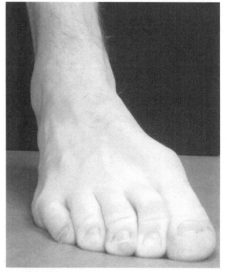

FIGURE 12B

Figure 12A shows the foot in its ideal position, called subtalar neutral, while figure 12B shows supination, and 12C pronation.

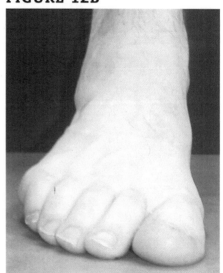

FIGURE 12C

As you have just seen, our feet are the ultimate foundation of your entire framework. They hold and support all of your weight-bearing joints, and will do so under incredible stress for seemingly endless miles in your lifetime. The data I have previously discussed show the critically important, documented role that structural alignment plays in arthritis.

I like bottom lines…so here it is…*Forty years of experience, constant trial and error, and subsequent treatment successes, has led me to redefine the cause of osteoarthritis of the weight-bearing joints as…excessive pronation occurring at the subtalar joint.*

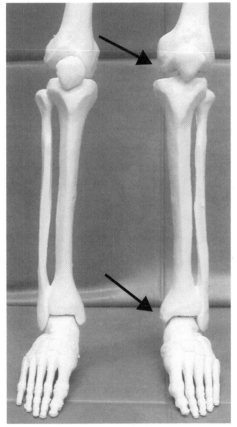 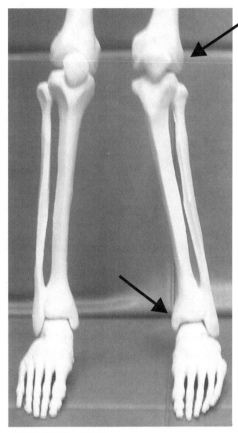

FIGURE 13A   **FIGURE 13B**

A properly aligned foot keeps the knee and all the other weight-bearing joints aligned, as seen on the skeleton's left side in Figure 13A. As soon as the foot begins to excessively pronate, the knee and all of the other weight bearing joints become unstable, as seen on the left leg of the skeleton in Figure 13B.

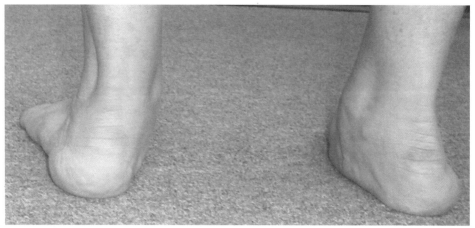

FIGURE 14B

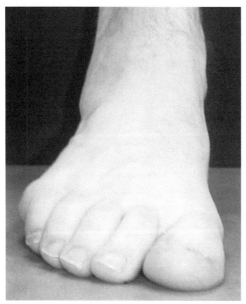

FIGURE 12C

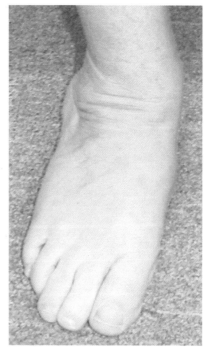

FIGURE 14A

There is a very important difference between the normal amount of pronation which is necessary to function properly during walking and running, as seen in Figure 12C, and the excessive, pathological amount of motion that causes arthritis of the weight bearing joints, as seen in Figures 14A and B.

Is it the only factor? Absolutely not. ..just the major one. We function as one integrated, kinetic chain. So problems occurring further up the chain can also certainly influence what is happening below. For example, tight hip muscles and curvature of the spine can cause pronation of the feet. And foot abnormalities of the opposite type…excessive supination or a rolling outward…can also cause arthritic problems of our weight-bearing joints too.

As you have seen in this chapter, the most powerful influence on our weight-bearing joints comes from our foundation which is our feet, and most specifically the subtalar joint. But how can so much joint damage be caused by one small joint in the foot? Granted…it sounds too simplistic. Well, read on, and I'll show you. You will also see how precisely controlling this joint can improve sports performance.

Remember, if you want to decrease abnormal stress on the weight-bearing joints, the first place to look…not the last…is your feet!

"*I was diagnosed with Juvenile Rheumatoid Arthritis at age 3, so throughout my life I have always had difficulty walking and running because of severe pain. I had unsuccessful major surgery on my feet. My rheumatologist, Dr. Gary Botstein who was trained at Harvard, sent me to Dr. Pack. Amazingly, Dr. Pack was able to correct my feet and ankles without further surgery and by so doing, take the stress off my knees and hips. Now physical activities are much easier for me. I can also work on my feet all day. I am so grateful to Dr. Pack for all that he has done for me, vastly improving my quality of life.*"

Lindsey Crumbley, Atlanta, GA

CHAPTER 5

Pronation and its Destructive Effects

*"As a patient of Dr. Pack's myself, and through the patients
I have sent to him, I can tell you first hand, that his ability
to properly and optimally position the foot, does indeed have
a great beneficial effect on reducing painful symptoms of the
knees and other weight bearing joints in those who have
arthritis "*

Gloria Singleton Gaston, MD, FACR
Harvard trained, board certified rheumatologist

- Excessive pronation is the underlying pathological process responsible for the destructive joint changes seen in osteoarthritis.
- Pronation causes a large number of problems within the foot, often thought to be due to age.
- Evaluation and treatment of excessive pronation in children can slow, and sometimes prevent osteoarthritis.

WHERE WE ARE...

Thus far we have dispelled the myth that age and weight...but instead... structural abnormalities, are the primary causative factors in the development of osteoarthritis of the weight-bearing joints. Much credible data has substantiated this. You have seen the importance of our feet...the neglected foundation of our entire skeletal system, and begun to realize that within the foot, the subtalar joint is the *structural core* of our entire body.

When excessive pronation occurs at this joint, the stage is set for extensive destruction, because all of our weight bearing joints become unstable and malaligned.

I have explained that pronation is basically a rolling in or flattening of the foot. But it is much more than that. And it is important that you understand pronation further, so you can see the specific effects it has on your joints when excessive. This will be discussed below.

Then, in the next chapter, you will learn what causes this destructive motion. Remember again, when I refer to pronation, I am referring to *excessive* pronation and it's powerfully negative impact on our system, and not normal pronation...some degree of which we all need.

PRONATION GOING DEEPER

Prior to reading this book, you may not have been familiar with the term *pronation,* or appreciate how significant this condition can be, but you most certainly have noticed people who have flat feet.

The word *pronation* comes from the Latin noun *pronus* meaning "bent forward." *Merriam-Webster's Collegiate Dictionary* defines *pronation* as, "rotation of the medial bones in the midtarsal region of the foot inward and downward so that in walking the foot tends to come down on its inner margin" [1, p. 994]; a bit wordy for me. More simply put, as we saw in Chapter 4, pronation occurs at the subtalar joint and results in a flattening of the medial or inside arch of the foot.

Pronation is like cholesterol. Although usually associated with problems, as I have mentioned, a certain amount of it is necessary for proper functioning. Not only does pronation provide the mobility our feet need to adapt to uneven surfaces, it also decreases some of the impact caused by walking. Certainly, not as deadly as high cholesterol level can be, excessive pronation can none the less, wreak havoc on our skeletal system and ultimately disable us.

---

3   Once again, as a reminder for clarity...whenever I use the term pronation...I am referring to an excessive, abnormal amount of motion.

Because of the powerful, deforming forces that excessive pronation inflicts on the skeletal system, it is the cornerstone of a podiatrist's practice. Pronation is as commonly seen by the podiatrist as cavities are by a dentist.

SELF-ASSESSMENT
Do You Pronate Excessively?

In rare instances, some people have a rigid, flattened foot whether they are sitting or standing. Most people do, however, have some arch at rest. Although you may need some assistance, you can easily appreciate this by doing a simple test. Sit in a straight chair with your shoes and socks removed while making sure that your feet can easily reach the ground. Place your knees directly over your feet, so that you are sitting with your upper and lower legs at ninety degrees to each other, as shown in Figure 15A

Without moving your knees, lean your feet slightly to the outside by

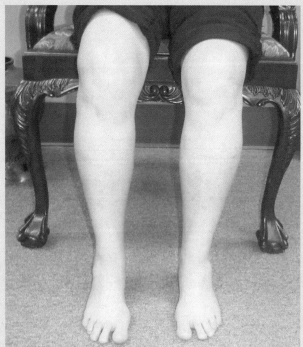

FIGURE 15A
Proper sitting position with feet on the floor
and your knees directly over your feet

lifting the insides of your feet off the ground… just enough so that there is no pressure on the inner edges of your feet as demonstrated in Figure 15b. Now notice the height of your arch. Although not possible in very severe cases, if properly positioned, an imaginary line down the middle of the front of your lower leg would go through your second toe.

Now without moving the location of your feet, stand up and let them assume a relaxed position. In most people, it can readily be seen that the arch lowers and flattens to some degree. The difference between what you have observed in Figures 15B and C is the effect of excessive, pathological pronation.

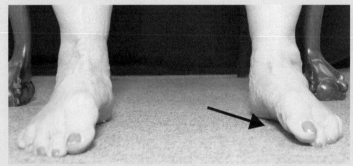

Figure 15B Leaning your feet slightly to the outside without moving your knees, so there is no pressure on the inner sides of your feet.

FIGURE 15B

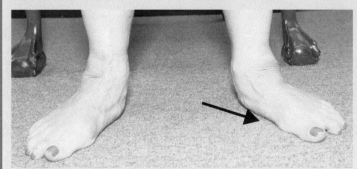

Figure 15C Standing, and noticing how much your arch collapsed.

FIGURE 15C

Clinically, the actual height of the arch is not as significant as the degree of change that occurs between sitting (non weight-bearing) and standing (weight-bearing). If an individual of average weight gets into a car and the car "bottoms out" or lowers significantly, there is a suspension problem. The same is true of you and the height of your arch. The more it *flattens* when weight is applied to it, the greater your suspension problem or structural instability.

THE IMPORTANCE OF WHAT YOU DON'T SEE

As mentioned above, the apparent lowering or flattening of the arch may seem to have no real significance to the untrained eye. This is especially true in the absence of symptoms such as arch fatigue or heel pain. Some people are even amused by the imprint their archless foot makes in the sand, or that their wet foot leaves on the ground. But this flattening is anything

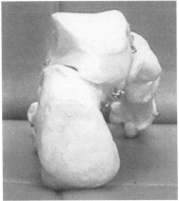

FIGURE 16A   **FIGURE 16B**

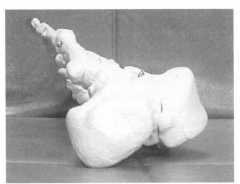

FIGURE 16C   **FIGURE 16D**

Figures 16A, B, C, and D, are all pictures of a foot skeleton as seen from the back. The top bone is the talus, and the bottom one, the calcaneus. Figure 16A shows optimal alignment and positioning of the talus and calcaneus. The remaining figures show what happens to these bones in the foot with excessive pronation. As the talus slips off of the calcaneus, the arch flattens as shown in Figure 16B. The forefoot or front part of the foot, then slides outward as shown in Figure16C, and rolls up and out as depicted in Figure16D. This can be seen quite easily in severe cases.

but innocuous. Like an iceberg, there is much more under the surface than what is apparent on the top. And like that iceberg, what is unseen…can be quite harmful. That's because flattening of the arch is only one part of pronation…a movement which actually involves motion occurring on three different planes…each of which can cause problems when excessive and specifically affect your arthritis.

As a matter of fact, these three components of pronation are so important, that there are actually different surgical procedures utilized to correct deformities on each of these planes, in very severe cases.

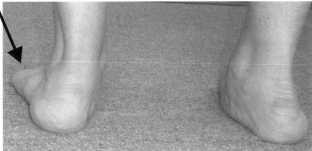

FIGURE 16E

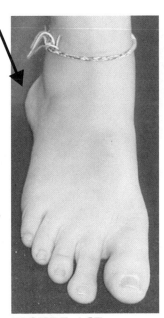

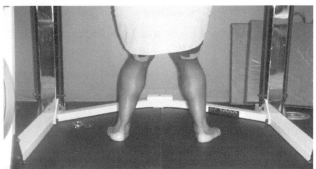

FIGURE 16F   **FIGURE 17**

Figure 16E shows a patient's right foot flattening significantly, and as it does, the front part of her left foot can be seen moving outward. Figure 16F shows the outward shift of the back part of another patient's right foot with pronation, which in severe cases, can even be observed from the front. Fig. 17, shows a professional weight lifter whose pronation is causing his feet to slide out from under his body when attempting to squat with heavy weights. Notice that his abnormal foot position is causing his knees to turn inward, placing a great deal of stress on them.

With pronation the subtalar joint "unlocks," and the foot loses its rigidity. Picture the two bones it comprises...the talus and the calcaneus...actually slipping out of position, much like two bricks in the foundation of a wall sliding off each other, causing the entire wall to become unstable. As a result, the foot rolls inward causing the arch to flatten, slides outward on the transverse plane (abduction), and continues an upward arc (eversion) (See Figures 16A, B, C, D, E and F). Excessive pronation actually causes your feet to slide out from underneath your body (See Figure 17).

In time, as this pathological process continues with each and every step taken, functional adaptation occurs. That is, continued abnormal joint motion changes the shapes of the bones, just as rubbing two rocks together over time will change their shapes. Functional adaptation thus creates changes in the joints that can prevent the foot from being able to assume its optimal, stable, functioning position. When this occurs, a previously far more easily treated, flexible flat foot can become a severely rigid, fixed deformity. This process occurs not only in the joints of the foot, but can affect any or all of the weight-bearing joints the foot supports. In time, changes occur that include the destruction of cartilage, bone spur formation, and joint-space narrowing... all signs consistent with the classical finding of osteoarthritis. This is precisely why it is so important to control excessive pronation in children... so as to avoid the potential for these often easily correctable problems to become major ones later in life.

Remember...
In most instances, it is this repeated sequence of excessive pronation, with resultant foundational foot instability, that causes excessive friction on the joints the foot supports. This underlying abnormal, mechanical process is responsible for the destructive joint changes that we call osteoarthritis.

Although I have mentioned this previously, because of current misconceptions, it warrants emphasizing again. Once abnormal mechanical factors are in place, age, additional weight and activity become important...not as currently considered primary factors...but as secondary ones to poor alignment. The longer one functions in an abnormal, poorly aligned position, or carries an increased load in that position, the greater the frictional pressure and subsequent damage to these joints.

The degree of severity of pronation can be observed clinically and confirmed with X-rays. Although rarely necessary, in very severe cases, surgery may be recommended to correct the deformity. In such instances angular measurements can be made from X-rays to help determine the specific procedures indicated. I did many of these over the course of my surgical practice…but only in the most severe cases. Now, as I have perfected conservative measures to control malalignment, I see the need for such surgery even less.

LOCAL EFFECTS OF EXCESSIVE PRONATION

In addition to causing sequential instability of all the weight-bearing joints supported by the foot, pronation can also cause typical arthritic signs and symptoms in areas of the foot itself.

Despite the all too common notion that some of these affected areas of the foot are isolated sites arbitrarily affected by osteoarthritis, and therefore thought to be mainly due to age and weight, they are instead, usually the result of specific mechanical factors, which when treated as such, can often lessen or eliminate symptoms. Sometimes this can help patients avoid the use of traditional arthritic medications or prevent the need for surgery. If utilized early enough, treatment measures that decrease stress forces may prevent arthritic problems from ever developing in the foot or elsewhere. This will be discussed further in Chapter 7.

That is not to say that some of these areas aren't sometimes affected by other types of arthritis…they are. In these cases the particular areas of involvement *are truly caused by a primary disease and not mechanics.* For example, psoriatic arthritis, and rheumatoid arthritis, may affect the ball of the foot.

But even in cases in which an actual disease process is the primary cause of weight-bearing joint arthritis, two other major factors are always at play. First, since as you know, we are not structurally perfect, patients will all have had some structural problems long before their disease occurred and was diagnosed. Second, their arthritis will over time have altered their bone and joint alignment, which affects the way they walk and function, thus causing additional problems with their other weight-bearing joints.

Because this concept is so important to your care, I believe it needs to be reiterated... *To achieve long-term positive results, joint problems cannot be medicated away, surgically removed, or optimally treated by physical therapy or any other means, unless the causative biomechanical issues are addressed first.* Because of this, an understanding of pronation and its causes and effects, are critically important to anyone offering treatment to patients with osteoarthritis or any other arthritic condition affecting the weight-bearing joints.

The examples that follow are some of the more common foot problems due to specific mechanical abnormalities. Since this is not a medical text, these conditions do not represent a comprehensive listing and are only briefly discussed.

BUNIONS AND TAILOR'S BUNIONS

Bunions are the bumps often seen at the big toe joints of the foot. They can also occur on the outside of the foot just behind the little toe where they are

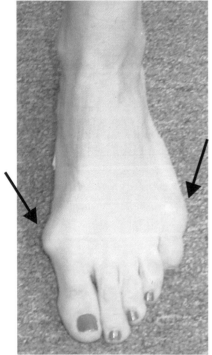

referred to as tailor's bunions (See Figure 18). In bygone times, seamstresses used the outside of their feet to run their sewing machines and developed irritations in this area, hence, the name.

Although most people think of bunions as just bony enlargements, most bunions involve some degree of dislocation of the big toe joint caused by pronation of the foot (See Figures 19A and B). Because of this, bunions are technically known as *hallux abductovalgus deformities.* Derived from Latin, this more accurately describes the dislocation, which consists of a rotation and movement of the big toe toward the other toes. If and when indicated, regardless of the type of surgical procedure used, it is also important to treat the pronation which caused the problem initially.

FIGURE 18
This patient has both a bunion and a tailor's bunion.

Although often considered to be a problem of older people, bunions can be found in young children. This is especially true if they have predisposing factors that make the big toe joint more prone to dislocate. These include being generally very flexible throughout the body, abnormal flattening of the foot and having a very round first metatarsal head, which is the long bone behind the big toe joint.

Many people think bunions are inherited, and that is true to some extent. However, what is really inherited is not the bunion, but the excessive pronation

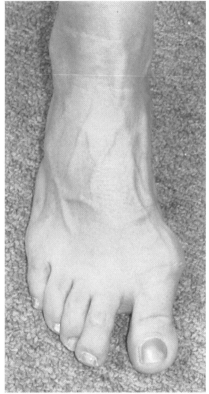

FIGURE 19A

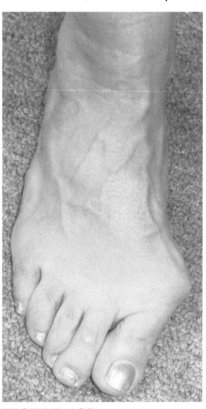

FIGURE 19B

Even though there is some bone enlargement, a dramatic difference can be seen in the patient's bunion deformity when she is sitting (See Figure 19A) and when standing (See Figure 19B). Notice how the big toe dislocates and shifts toward the second toe when the foot bears weight, and pronation occurs. It is important to note that the same destructive impact you see here…that pronation has on the big toe joint…it also has on the knees and other joints the foot supports.

or rolling in of the foot, which in turn, causes the bunion. So if you have this deformity, be sure to have your children evaluated for pronation. Like other joint problems discussed in this book, bunions can usually be prevented.

A common misconception is that bunions are caused by pointed footwear, like women's high-heeled shoes. However, many women who have worn such shoes for years do not have bunions or only have them on one foot. While others, who have never worn this type of shoe, have severe bunions. Shoes can exacerbate the problem, but in almost all cases, they are not its primary cause.

Left untreated, bunions can not only make buying and fitting shoes difficult, but may become very painful, disfiguring and cause problems in other joints. That's because as the dislocation progresses, the big toe may continue to shift laterally toward the second toe, pushing it up and creating a hammertoe (See Figure 21).

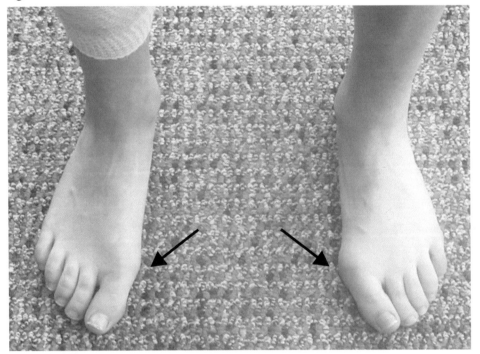

FIGURE 20
The nine year old girl seen in Figure 20 already has bunion deformities. Her left foot is much worse due to the far greater amount of pronation on that side.

Remember that having a bunion is not an independent, isolated condition. The same poor alignment that causes bunions also causes other weight-bearing joints to deteriorate. So, if you are developing signs of a bunion, even in the absence of symptoms elsewhere…such as knee pain…be sure to treat the excessive pronation. Early treatment can slow the progression of your bunions, help avoid other arthritic symptoms, and prevent a recurrence after surgery.

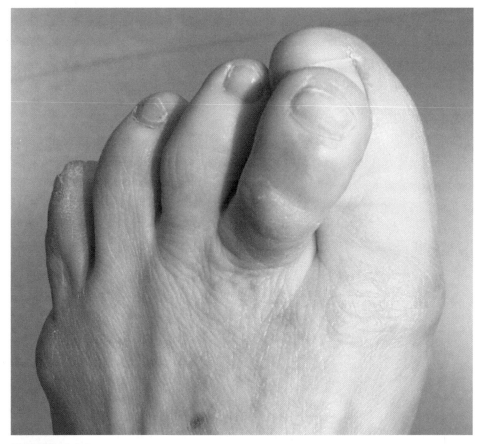

FIGURE 21

This patient has a hammertoe deformity of the second toe caused by her bunion.

[4] Although, as I've mentioned, these osteoarthritic problems of the big toe joint are due to mechanical factors, this joint can be affected by other arthritic disease processes, such as gout. As most cases of gout primarily affect the big toe joint, I believe it will eventually be shown that underlying joint stress caused by mechanical factors actually predispose this particular joint to gouty attacks.

HALLUX RIGIDUS (HALLUX LIMITUS)

As mentioned briefly above, the rounder the distal end of the first metatarsal (the long bone behind the big toe), the greater the potential for this joint to become dislocated on the transverse plane in those who pronate. That is why some people have very large bunions. Individuals, whose joints are more squared off at the ends, cannot dislocate them as easily. In these cases, pronation causes their joints to jam, and can result in far more severe arthritis with bone spurs, and a limited range of motion, yet without any dislocation. This condition is known as *hallux rigidus* or *hallux limitus*, because of the extremely limited motion that can occur at this joint. Like bunions and tailor's bunions, this condition can occur in one or both big toe joints. In many cases, patients with this condition eventually require a big toe joint replacement.

In its most severe form, this deformity typifies osteoarthritis with its narrowing of the joint space, bony proliferation, destruction of cartilage, cystic formation, achiness and soreness, and limited range of motion

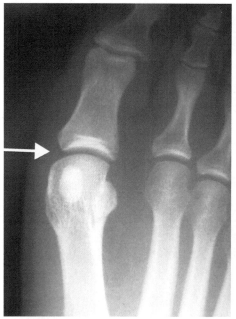 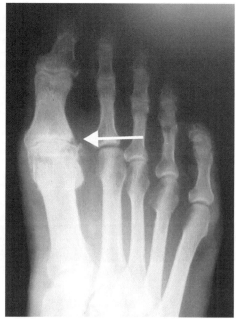

FIGURE 22   **FIGURE 23A**

Figure 22 is an X-ray showing a normal big toe joint space. The remaining X-rays show a patient with hallux limitus of both big toe joints. Notice in Figure 23A, the obliterated big toe joint space of this patient's right foot.

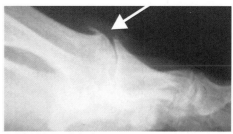

FIGURE 23C

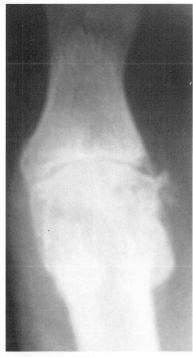

FIGURE 23B

Figure 23B is a close up of this deformity. Here you can see the obliteration of the joint space, especially centrally, as well as bone spurs on the outer right side of this joint. Notice again, the difference in this joint space from the normal space in Figure 22, as well as the difference in this patient's big toe joint and his unaffected second toe joint space. Figure 23C shows a lateral or side view of his right foot. Large bone spurs can be seen on the top of this joint. These spurs further limit motion. Figure 23D and e are X-rays of his left foot. Just as in Figure 23A, notice the obliteration of the big toe joint space, especially as compared to his second toe joint. A close up of this joint (Figure 23E), shows what appears to be complete closure of this joint with bone spurs seen on the outer left side of the joint.

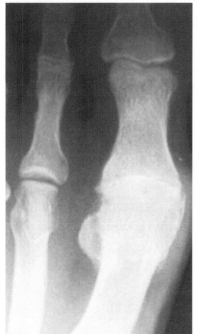

FIGURE 23D

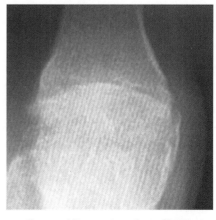

(See Figures 22, and 23A, B, C, D, and E). In these cases, it is most often assumed by many physicians unfamiliar with the pathomechanics involved, to be just another joint affected by the ageing process. But make no mistake... this condition is due specifically to the destructive forces of pronation on a square shaped, first metatarsal head. I have replaced these joints many times in young and middle aged individuals who had no history of injury, other joint symptoms, or any laboratory findings indicative of any arthritic disease whatsoever.

As mentioned above, arthritis of any weight bearing joint regardless of its cause, will over time affect others. Hallux rigidus is a prime example. Since patients with this condition cannot flex or bend their big toe joints, they usually walk on the outsides of their feet which will further alter their gait, possibly causing other foot problems, as well as painful symptoms in the knees, hips, and lower back. Similarly, patients like those with rheumatoid arthritis of the knees, whose disease has changed the way they walk and function, will have other weight-bearing joints affected, too. Altered gait patterns will always occur when a joint necessary for proper function is compromised.

HAMMERTOES, METATARSALGIA, AND METATARSAL PHALANGEAL (MP) JOINT PAIN

As you have seen above, hammertoes are contracted digits (toes) that sit up on the tops of the feet (See Figure 21). If you're not sure what that really means, simply place the palm of your hand on a flat table and bend or flex your fingers, pulling them back and upward while leaving much of your palm in place. Picture your toes rigidly fixed in this position and that is what a classic case of hammertoes looks like (See Figure 24).

Although hammertoes can start at an early age, people are rarely born with them. Quite flexible when they first appear, hammertoes usually become progressively more rigid. In the late stages they can become fixed as if fused into that position. The increased friction caused by the continuous rubbing of shoes on the tops of these toes often results in painful corns which are areas of hardened, protective skin. In severe cases these areas can form ulcers which can become secondarily infected.

As the toes continue to be pulled upward and back, they cause jamming of the joints to which they are attached. These are called the metatarsal phalangeal joints, and are located where the base of the toes or phalanges attaches to the long bones or metatarsals just behind them. This often results in pain and inflammation in this area on top of the foot.

If the process continues, hammertoes can exert a downward, retrograde force on the metatarsals, resulting in pain on the bottoms or balls of the feet. Symptoms are intensified, because the normal protective fat pad on the ball of the foot becomes displaced forward, exposing the metatarsal heads. The combination of increased bony pressure and lack of shock absorbing protection can make every step quite painful. This condition is called metatarsalgia. When necessary, the surgical correction for this type of contracture can be very extensive (See Figure 25).

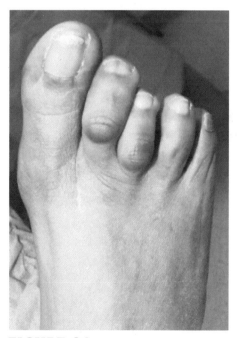

FIGURE 24

A classic case of fixed, rigid hammertoes.

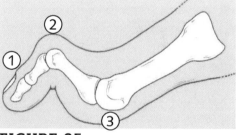

FIGURE 25

This diagram shows the sequence of events described above. This can result in a painful corn developing on top of the toe (1), jamming of the metatarsal phalangeal joint, with pain and inflammation in this area on top of the foot (2), as well as severe discomfort on the bottom of the metatarsal or ball of the foot (3).

To appreciate this painful mechanism repeat the simple exercise you did earlier, bending or flexing your fingers by pulling them upward. But this time don't place them on a table. Look at the palm of your hand when you do this. Pay particular attention to the area just behind where your fingers attach to the palm of your hand, and you will readily see the protective fat pad move forward and upward. Feel the bones under the fat pad and notice how much more prominent or exposed they become when the pad is displaced.

A number of conditions can cause hammertoes. These include arthritic and neurological diseases such as rheumatoid arthritis and cerebral palsy, as well as certain types of injuries. When hammertoes are diagnosed as osteoarthritis, they are usually associated with very tight calf muscles. This in turn causes the mechanical imbalances that result in these problematic toes.

PLANTAR FASCIITIS AND HEEL SPUR SYNDROME

The fascia is a band of tissue under the arch on the plantar or bottom of the foot. In most cases pain in the plantar area is mechanically induced and often precipitated by tight calf muscles, which cause excessive pronation and a subsequent stretching of the fascia.

Symptoms usually occur gradually and are most commonly described as a localized achiness or soreness. If this condition is allowed to continue, the excessive pull of the fascia at its attachment at the bottom of the heel, can produce a bursitis or inflamed sac in this area, with or without bony proliferation. The fascia may also rupture causing sudden, severe pain. If the pain is predominantly in the arch, it is called plantar fasciitis, whereas pain in the heel is referred to as a heel spur syndrome.

In many cases significant heel pain is common and at times disabling. Generally, symptoms are greatest upon arising in the morning or after sitting for extended periods of time. Although patients often logically think that their pain should actually be less after a night's rest, it is typically worse. This is due to fluid build-up in the bursal sac under the heel. While walking causes the bursa to become inflamed, it also pumps the fluid out. When ambulation ceases, fluid accumulates and until the first few steps are taken forcing the fluid back out, the pain can be quite severe.

Symptoms may not be consistent with the degree of pronation or the radio-graphic signs of spur formation. Patients with only mild pronation and no spur formation may experience pain, whereas others with very large spurs and severe pronation may have little discomfort. The same is true regarding the correlation of weight. Although too much weight can certainly make this condition worse, it often occurs in people who are not overweight at all. When a heel spur syndrome is conservatively treated with such means as injection therapy, orthotics and appropriate stretching, surgery can almost always be avoided.

Pain under the arch and heel are very common in people who are actively in-volved in sports or who do any activity in which the fascia is overly stretched… such as standing on a ladder or squatting when gardening. As with the other conditions described in this section, pain in these areas can also be due to causes other than those that are mechanical. These include primary arthritic conditions such as rheumatoid arthritis and Reiter's syndrome (See Figure 26).

RETROCALCANEAL BURSITIS AND HAGLUND'S DEFORMITY

Painful symptoms in the back of the heel or retrocalcaneal area are quite common. Signs and symptoms including bursitis, possible tendon rupture, and large spurs called Haglund's Deformity can develop (See Figure 27A, B). This is due to the continued pull of tightened calf muscles where they attach to the back of the heel by way of the Achilles tendon. Other arthritic conditions such as gout can cause inflammation in this area in the absence of mechanical factors. Tightened calf muscles are almost always a factor, especially when bone spurs in the back of the heel are present.

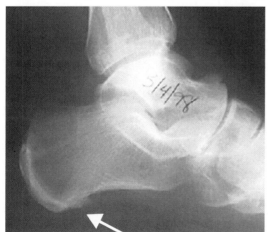

FIGURE 26

Figure 26 shows a spur on the bottom of this patient's heel, due to pronation and the subsequent pull on the plantar fascia.

SINUS TARSITIS

As previously discussed, the subtalar joint consists of the talus which attaches to the bones of the ankle, and the calcaneus or heel bone. A canal or groove called the sinus tarsi, runs obliquely through the subtalar joint and contains a neurovascular bundle, which is a complex of nerves and blood vessels. It is not uncommon for this area to become inflamed with excessive pronation and result in a condition called sinus tarsitis. As I have mentioned before, because of the close proximity of the subtalar joint to the ankle, pain in the former is sometimes misdiagnosed as ankle pain. This is most easily differentiated by pressing directly over the subtalar joint laterally, while inverting the foot (rotating or rolling it inward).

METATARSAL CUNEIFORM EXOSTOSIS

Pain on the top, middle of the foot, in the area of the metatarsal cuneiform joints, is quite common and often diagnosed as osteoarthritis. Symptoms are consistently seen in patients with a very high arched or cavus foot as well as in those with pronation.

Increased friction in this area results in spurs that can cause significant discomfort during walking (See Figure 28). This is because the spurs lie directly

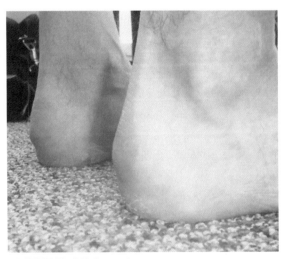
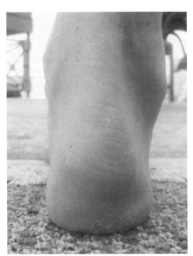

FIGURE 27A   **FIGURE 27B**

Large spurs on the backs of the heels are seen in this patient with Haglund's Deformity.

beneath the area where shoe laces are often tied. A small nerve runs directly over these joints. When spurs are present, they push this nerve upward making it more susceptible to the downward pressure created when shoes are fastened. A nerve entrapment or compression syndrome can develop. This can produce symptoms which include sharp, radiating pain, numbness, and tingling. Because of this, some patients are misdiagnosed as having a neurological disease such as peripheral neuropathy. Others may develop a bursitis such as that described above.

POSTERIOR TIBIAL TENDON DYSFUNCTION (PTTD SYNDROME)

The posterior tibial muscle primarily attaches to the inside and bottom of the arch via a tendon of the same name. It is the primary force counteracting pronation. When pronation occurs, this tendon can be pulled and stretched at its bony attachment in the same manner as the heel spur syndrome described previously. Referred to as insertional tendonitis, its symptoms can be quite severe. With prolonged stress and/or extra weight, this tendon can pull fragments of bone off its primary bony attachment and may even rupture spontaneously, resulting in a sudden, very flattened foot (See Figure 29).

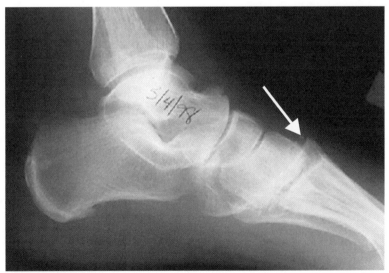

FIGURE 28

A metatarsal cuneiform exostosis or spur, as described above is seen on this lateral X-ray.

NIPPING PRONATION IN THE BUD

Child obesity is of great concern to physicians because of the subsequent development of such health issues as arterial plaque, diabetes, and high blood pressure that may occur as children age. Because it is the primary cause of arthritis of the weight-bearing joints, pronation should warrant the same degree of attention in children.

Unfortunately, pediatricians, rheumatologists, internists and physicians in many other specialties, often pay little attention to flat feet. This is due largely to the time constraints placed on them as they deal with other, sometimes more urgent medical issues...and frankly...because the real significance of this condition is often not fully appreciated.

If abnormal flattening of the feet *is* noticed, pediatricians generally tell parents, "not to worry," because the child, "will most likely outgrow it." Part of the reason for this explanation is that children often have a natural fat pad on the bottoms of their feet that obscures the arch and will usually disappear. This, however, should not be confused with pronation, which causes an actual flattening of the foot. As mentioned earlier, in such cases instead of *outgrowing* their flexible, usually easily correctable deformities, functional adaptation develops over time, and children often *grow into* far more serious, fixed, rigid problems by adulthood. These problems are often accompanied by painful symptoms and destructive joint changes...most of which are preventable.

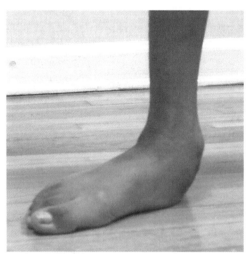

FIGURE 29

The sudden onset of a very painful, flattened foot in a patient with a ruptured posterior tibial tendon.

To keep your child's structural core stable, pronation should be controlled as soon as it is diagnosed. It is therefore important to have your children's feet checked regularly, and their skeletal structure properly evaluated by a podiatrist or a physician well versed in children's foot problems. Indeed, "cautious waiting" can have disastrous results on the foot and all the other weight bearing joints it supports.

"*Dr. Pack,*

I spent the weekend with mother after you treated her. What a change!! She needs no help walking or stepping off curbs or getting out of the car. Although she has had severe arthritis for many years, she now has no pain in her knees or in her back. She said she never dreamed that she would be pain free again in this lifetime. She vacuumed and mopped last week — in one day — again something she never thought she would be able to do again. Saturday, we went to Athens and I came back exhausted. And at 86 she was still going — like the Energizer rabbit! How can we ever thank you enough for adding quality to the quantity of her days on this earth."

"Diane Huff, Union Point, GA

CHAPTER 6

Two Major Causes of Pronation

"*I became aware of Dr. Pack as a Rheumatology Fellow at the Medical College of Georgia, when one of my professors invited him to speak to our staff. The physician, also a rheumatologist, was a middle aged runner and had significant problems that Dr. Pack was able to help him with. I have been quite impressed with the immediate increased improvements in balance, alignment and stability as well as the changes in position of the weight bearing joints that can be created by his techniques. Our staff has gained a great deal from his teachings and I know that his work would be a great asset to patients if implemented into mainstream medicine.*"

Alok Sachdeva, MD, FACR

Board certified rheumatologist, Former Assistant Professor of Medicine, Department of Rheumatology, Medical College of Georgia

- Leg-length discrepancies and tight calf muscles are two of the most common, yet often untreated causes of excessive pronation.
- Unequal leg-lengths can be structural or functional.
- Abnormal alignment and tight muscles can restrict and cause joint pain.
- Tight calf muscles are usually caused by improper stretching and the body's protective mechanisms of splinting and guarding.
- Proper stretching can eliminate most of the problems associated with tight calf muscles.

As you have seen, there are a number of conditions that are *caused* by pronation. These include problems in the foot as well as the weight bearing joints the foot supports. Because of the significant role it plays in arthritis, it is also important to understand the underlying conditions that are the *causes of* excessive pronation.

Anatomical variations in the bones of the foot are very common. For example, the abnormal positioning of the five metatarsals of the forefoot can cause pronation. This is known as a forefoot varus deformity and is typically due to a birth defect. Rotational problems, such as a rearfoot varus deformity, which is a turning of the calcaneus or heel bone in an outward direction, can also be a cause. Children and adults, who are excessively flexible, are also more prone to excessive pronation because of the inherent instability their loose ligaments allow. Skeletal, muscular, and neurological abnormalities in the upper and lower leg, scoliosis, and injuries are among the other causal factors.

Two of the most common and powerful deforming forces of pronation that can often be easily corrected, but are generally overlooked, are leg length discrepancies and tight calf muscles. Understanding the causes and effects, that they have on the body's structure, and therefore on arthritis of the weight-bearing joints, can help you live a more pain free and active life.

LEG-LENGTH DISCREPANCIES

I was asked to go to California to see a very promising 9-year-old, competitive golfer. In less than 20 minutes after treating her, she was able to go from hitting the ball 141 yards, to more than 180 yards off the tee. Subsequently, she came to my Atlanta office and with some additional fine tuning hit a ball 210 yards. Granted…she did not do this consistently…and not everyone experiences this degree of improvement. But I mention it here, because it illustrates the powerful and immediate changes…the untapped "edge"…that often can be created with athletes. There will be more about increasing golf and other types of athletic performance in chapter 8.

What was the apparent magic that caused such a dramatic change in her sports performance? It was the same, usually neglected treatment that can prevent and perhaps dramatically improve your arthritic symptoms…limb length equalization.

DEFINITIONS, CAUSES, AND IMPLICATIONS

As the name implies, a limb length discrepancy is a difference in length between one's right and left legs. Practitioners disagree about how widespread or clinically significant this discrepancy is, whether or not current methods of assessment are either reliable or valid [1], and if and when to treat it. However, the general consensus is that a difference of more than a quarter of an inch is abnormal.

Leg-length differences are not usually considered significant in osteoarthritis. As a matter of fact…they are generally not considered important in most aspects of medicine or in sports. Because of this, they are rarely checked. If and when significant differences are found, treatment is usually nonexistent or incomplete. But a study conducted as part of the Johnson County Osteoarthritis Project, which is an ongoing study sponsored by the CDC, the NIH, and other entities, impressively showed that *patients who had a significant leg-length discrepancy, were found to have almost twice as much evidence of arthritic change on X-rays than those who had none. They were also 40% more likely to have symptoms [2].*

Hip, knee, back pain and stress fractures have been associated with leg-length discrepancies. Significant improvements or total alleviations of back pain have been reported when leg length differences were corrected [3]. Another study performed at the Texas Scottish Rite Hospital for Children in Dallas, Song et al. [4] showed that more mechanical work was handled by the longer leg and that the center of body mass was displaced more in children with leg length discrepancies.

Leg-length discrepancies effect far more than just our mechanics. Gurney and coworkers [5], for example, reporting in the *Journal of Bone and Joint Surgery* in 2001, found that oxygen consumption and perceived exertion were increased in the patients involved in their study who presented with leg-length issues. They also stated that elderly patients with significant medical problems may have difficulty walking, even with small leg length differences. Muscle fatigue was also noted.

[5] Technically speaking, because we are not born symmetrical or perfect, everyone has a structural leg length discrepancy to some extent. In many of these cases this may not be clinically significant and not need treatment, whereas others are, and should be corrected. In the same way I refer to pronation, I now refer to leg-length discrepancies…those cases that are excessive, clinically significant, and need treatment.

As I stated earlier, leg-length differences are also generally not considered important and are not accurately dealt with before or after hip replacement surgery. Konyves and Bannister [6] reported that in their study, that 71% of the hip replacements were performed on the shortened side, and that post-operatively, nearly as many had been lengthened. *From these statistics, you can easily see why one's mechanics would be significantly altered from such surgery.* Others agree and have shown that postoperative leg-length discrepancies are indeed quite common and can be significant at times [7, 8].

Despite the general lack of appreciation for the important role that leg length discrepancies play in osteoarthritis, my experience has shown that the treatment of this condition often has a very positive and sometimes immediate effect on arthritic symptoms in the weight-bearing joints. So I will talk more about this later in the chapter.

STRUCTURAL VS. FUNCTIONAL DISCREPANCIES

Limb length discrepancies are commonly grouped into two categories: structural and functional or positional. Everyone has a structural leg-length discrepancy to some degree, just as we may have one arm longer or foot flatter than the other .

A structural difference means that one or more bones in one leg are actually longer or shorter than those in the other leg. The affected bone(s) can be the upper leg bone called the femur, or the lower ones, which are the tibia and fibula. Structural leg length differences are most often due to congenital causes...meaning they were present at birth. Any condition that affects the growth centers in children can cause permanent shortening on the affected side. These include bone infections, fractures, and neurological conditions like polio.

A great many of us also have functional differences, which are those due to tight muscles or other causes. Functional or positional limb-length discrepancies occur as compensatory mechanisms for problems elsewhere. For example, certain activities like swinging a golf club or tennis racket, can cause the pelvis to slip up on one side. A painful arthritic problem in one joint like the hip could also trigger protective muscle spasm and guarding and result in a shortening on the side that is involved. Indeed, in all my years of practice,

I've never examined anyone who was absolutely symmetrical. Some have had differences greater than an inch (2.5 cm) and had never been aware of it, while others knew they had much smaller differences and never had it treated.

As you will see shortly, it is important to first determine which type of discrepancy is present and then treat it accordingly. *Failure to do this can create additional problems rather than alleviate them.*

IMPLICATIONS

Regardless of the underlying cause, as you have seen, excessive foot pronation can result in malalignment all the way up the weight-bearing, skeletal chain. Appreciation for this mechanism has been reinforced to me over the years by the often dramatic relief of symptoms in the knees, hips, back, and neck, once foot alignment is established.

As the foot pronates or flattens, the tibia (inside lower leg bone) turns inward while the femur (upper leg bone) turns outward. This causes stress on the inside of the knee joint. The turning of the upper and lower leg in opposite directions is similar to the twisting motion created by wringing out a washcloth. Destructive, frictional forces can thus be created on both the knee and hip. It may also cause *genu valgum,* or a knock-knee condition, in which the knee turns inward. Although being knock-kneed can have other causes, there is evidence that arthritis of the knee can occur when this condition is due to pronation.

As the skeletal system continues to become malaligned, a wide variety of problems can occur in the weight-bearing joints above the hip...the back, spine, and neck. This occurs specifically as a consequence of the sequential, mechanical malalignment originally initiated by excessive foot pronation. These problems are not limited to alterations in the joints, but can involve tendons, muscles, ligaments, and the other soft tissue structures that support and attach to these joints. Symptoms are dependent upon which particular structures are involved and to what degree.

We are one functioning, kinetic chain. As mentioned previously, pathological problems higher up the chain like scoliosis can cause pronation. Similarly, pronation can be caused by structural problems within the foot itself which will alter the function of the joints above it.

Regardless of which way the pathological process occurs... the bottom line is... how your foot is functioning. Excessive pronation must be controlled. Once again, the greater the degree of abnormality, the more active you are and weigh, and the longer you function in this abnormal position, the greater the degree of degenerative arthritis you are likely to have.

Understanding this basic premise should help you appreciate why it is so important that malalignment be evaluated and corrected in all patients with arthritis of any kind that affects the weight-bearing joints.

DIAGNOSING AND TREATING A LEG-LENGTH DISCREPANCY

DIAGNOSIS

Many patients have never really thought much about having a leg-length discrepancy, or have appreciated its importance. When questioned however, the presence of this condition is often revealed. Signs and symptoms include having a noticeable difference on each side when getting pants hemmed, favoring one leg more than the other, or having difficulty in standing for long periods of time. A head and shoulder tilt or an apparent difference in arm lengths, which can easily be seen by simply observing how a patient stands and walks...something most physicians rarely do...may also indicate this problem. Other patients report that their chiropractors always tell them that one leg is longer than the other, but often nothing is done to *permanently* correct this. It is important to note that other conditions like scoliosis can cause similar findings.

Perhaps most impressive, are the changes that occur in the lower extremities of individuals with this condition. Consistently, patients with a leg-length discrepancy stand with one knee more flexed or bent, and one foot more pronated or flattened than the other (See Figure 30A).

I have found that with small differences in leg-lengths, unknowingly, patients will often bend the knee and flatten the foot more on the longer side in an attempt to shorten it, while standing on the outside of their other foot as they try to increase the length on that shorter side. When leg-length differences are excessive, patients will often have greater flattening on the shorter side, since that is the only way that the foot can reach the ground. This is frequently seen in cases like polio where significant shortening can occur.

Even while sitting, a certain position is usually assumed by those with leg-length differences that are significant (See Figure 30B).

THE PHYSICAL EXAMINATION

Leg-length differences are most readily detected during a physical examination by a physician, physical therapist or other professional, while the patient lies flat on his or her back and tries to straighten and fully extend both legs. The examiner can readily see discrepancies by looking at the heels of both feet (See Figure 30C). Often a knee is contracted or bent from arthritis or other causes and cannot be fully straightened. In such instances, that leg is obviously longer than it appears. This part of the examination does not help the practitioner determine exactly where the leg-length difference is, the exact degree of difference, or the cause or type of problem present. A general screening of this nature does, however, provide a good starting point for determining *if* a problem exists.

The most common way of detecting leg-length discrepancies is by taking measurements from the hips, and also from the umbilicus, to the insides of the ankles. Methods such as sophisticated video recordings of gait analysis and pressure mapping, which utilizes sensors to compare the pressure from each foot while the patient is walking, can also, be helpful. If the pressure map shows a

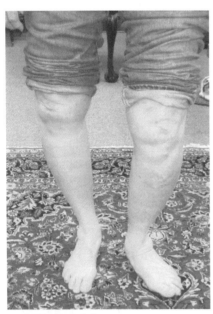

FIGURE 30A

Notice the bent left knee and more flattened foot on this patient's left side due to her longer left leg.

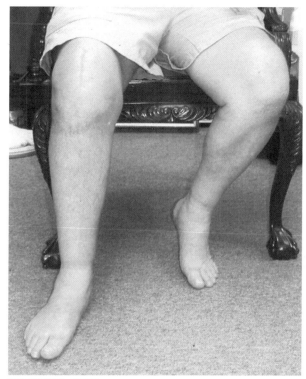

This patient has a much longer right leg as a consequence of joint replacement surgery on his right hip. Patients with this problem will usually place their right foot out, further away from their body. and roll their foot inward or outward as they try to shorten that longer leg. While doing this, they will most often keep their shorter leg closer to them and more flexed or bent, while putting pressure on the ball of that shorted foot in an attempt to raise it. This, they find is a much more comfortable sitting position. Notice this patient is doing exactly that.

FIGURE 30B

Notice this patient's significantly longer left leg as a result of hip joint replacement surgery on that side.

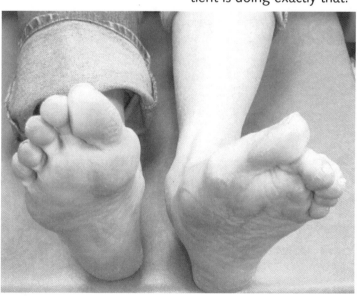

FIGURE 30C

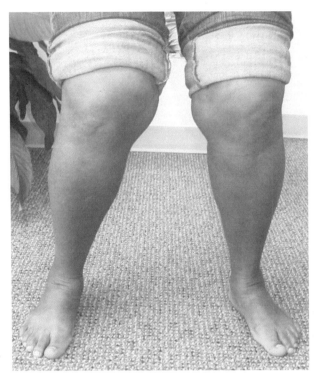

FIGURE 31A

The difference in vertical alignment of this patient's right knee can readily be seen when squatting, without (Figure 31A) and with temporary corrections (Figure 31B).

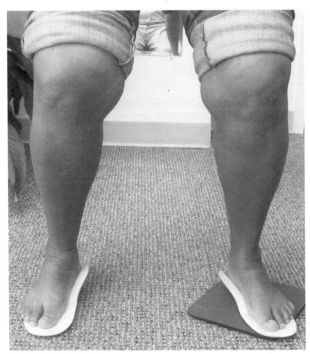

FIGURE 31B

difference in the force or pressure exerted by each foot at different phases of gait, a limb-length difference is likely. Traditional X-rays can also be utilized, as well as more sophisticated ones, like scanograms.

One of the simplest and most impressive diagnostic tools a practitioner can use when evaluating a patient is simple lifts. These can be small platforms of wood, hardened rubber, or other materials large enough to fit the whole foot. When asked to perform a partial squat, the patient who pronates excessively will usually turn one or both knees inward. This occurs more often, but not always, on the longer side.

I try lifts of different sizes in quarter-inch increments and place them under the shorter leg. With each additional correction, the patient is asked to squat again. Typically, the squat will be easier as optimal correction is reached. Although some patients cannot discern improvements, most will generally feel an increase in balance and stability which the lift provides, and most importantly... experience less pain in an arthritic knee... and sometimes in the hip, or back. More vertical alignment of the knee will also be evident. Positive findings such as these are a good indication that incorporating a lift as part of the treatment will be helpful in eliminating painful symptoms (See Figures 31A, B).

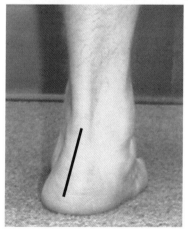
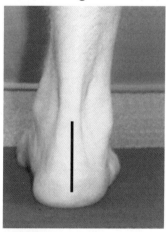

FIGURE 32A   **FIGURE 32B**

In Figure 32A, the patient is standing on the outside of his right heel in an attempt to lengthen his shorter right leg. Notice how much more vertically aligned his right heel is when he is standing on a ¼" lift (Figure 32B).

SELF-ASSESSMENT

DO YOU HAVE A LEG-LENGTH DISCREPANCY?

1. Lie flat on your back with your feet and legs as straight as possible. Ask someone to help you and have them put their hands around the backs of your ankles. While they are lifting your legs slightly off the ground, they should pull your legs toward themselves a couple of times. Then while you remain still, keeping your ankles and feet together, have them place both your feet and legs back down on the flat surface. Looking straight down, they should be able to see whether your heels are even, or if one is closer to them than the other (See Figure 30C).

2. Preferably, while standing in front of a mirror and holding on to a railing or wall for support, or perhaps having someone hold your hands, try to do a partial squat. Be sure to keep your heels flat on the ground and don't worry about how far down you can go. Notice whether one knee turns inward and hurts more when you do this. Also sense how balanced or unbalanced you feel. Then place a small ¼ inch thick book or other object under the foot on the side you think is shorter and repeat the squat. Be sure to choose something to stand on that isn't slippery. For additional confirmation, you can try the lift under the other side and see which feels better. As I have mentioned, some people will immediately have less pain in their knees or other weight bearing joints, and also feel more balanced while trying this, while others may not really feel much of a difference (See figure 31A, and B).

3. Stand with your feet shoulder width apart. While keeping your feet in place and your arms straight out from your sides, turn at the waist to your right side and then to the left. Take a visual picture of something on the wall that you were able to see when fully turned. Because a longer leg makes it more difficult to turn to that side, putting a lift under the short leg will often increase your ability to turn to the opposite side.

Again, by offering these self-assessments, I am not implying that you should treat yourself with a lift in your shoe or by any other means without first checking with your physician. You could, for example, put the lift on the wrong side and cause more problems. Since many physicians do not check for conditions like this, *these self-assessments are only meant to help you determine whether you have a leg-length problem, how significant that problem may be, and if so, direct you to getting the professional help you need.*

I have never ceased to be amazed at the dramatic and immediate effect these simple tests often have. Even in severe cases of osteoarthritis of the hips and knees in elderly patients, symptoms can often be reduced. This is especially true when leg-length differences are greater than a quarter of an inch...which is almost always the case with joint replacements.

As mentioned, patients with a short leg will often stand on the outside of that foot in an attempt to raise their shortened leg. These patients can often feel an immediate change in the position of that foot...especially in their heel... when a lift is placed under the shorter leg, even if they have never noticed the problem before. If they look for it, experienced physicians can readily notice this difference (See Figure 32A, B).

I always have the patient stand on sample foot inserts in addition to the lifts just described (See Figure 31B). By adding corrections for excessive pronation to determinations made for any leg-length discrepancy, far better alignment can be created and a further reduction of symptoms noted. Chapter 7 contains a more detailed discussion of this evaluation and treatment protocol.

You can try the following self-assessments at home, to see whether or not you have a leg-length issue. *Remember, I certainly am not implying that you can accurately diagnose your own problems, nor am I suggesting in any way that you treat yourself.* These simple tests can however often help you determine if you do have a problem and *how extensive* it may be.

TREATMENT

In many ways, medicine is still an art, and treating leg-length discrepancies is a great example of this. Determining that someone has this problem is easy, because the clinical signs and symptoms are often obvious. It is far more difficult to diagnose whether the discrepancy is structural, functional, or both, determine the exact degree of discrepancy and decide how much correction the patient can tolerate. For this, I often refer patients to a really good physical therapist or kinesiologist. It is worth noting that some of these specialists are far more experienced in assessing this type of problem than others. Sometimes even the best can't be sure.

The accurate diagnosis and treatment of a leg-length discrepancy poses many challenges. As mentioned earlier, if you have an arthritic right knee, chances are you cannot fully straighten it. This might cause your right leg to appear shorter, although that may not really be the case at all. In such instances a lift might be used on the shorter side. Then, if physical therapy enables you to increase the range of motion of your right knee, the length of that leg may be increased. If you continue to wear the same lift you will be preventing your knee from fully straightening, thus making it worse.

The same is true with various other compensatory mechanisms. For example, if an individual with a structurally longer right leg is flattening their foot on that side in an attempt to shorten it, and a foot orthotic is made that decreases that pronation, treatment will actually have lengthened that leg. As a result, back pain and other problems may occur or worsen.

People who are shorter and more active are likely to have symptoms caused by smaller differences in leg lengths. So a difference of a quarter of an inch affects a 5-foot-tall golfer and a 6-foot-tall sedentary attorney quite differently. This is especially true for someone who has had an injury or is overweight, because of the increased pressure on the joints.

As I have mentioned, besides the structural differences we all have, many of us will also develop functional discrepancies. This is due to the many physical activities we have all participated in during our lifetimes. Unlike structural leg length differences, which are commonly due to birth abnormalities and do not change unless an injury, surgery, or other complication occurs, functional

leg-length differences can change. Such changes occur for a number of reasons. For example, swinging a bat or golf club hard or experiencing a running
injury, can cause you to "pull" a muscle in one of your hips, lower back or
other area. A sudden turn or lifting a heavy object can have the same effect.
As the body responds with muscle tightening and guarding on the affected
side, a shortening of that leg may occur.

I often see professional athletes, whose leg lengths periodically change from
side to side, corresponding to the frequent injuries they may encounter. This
has a very significant effect on their performance. Just as remarkably, I have
also seen these problematic, functional leg length discrepancies change...at
times immediately...with professional, corrective manipulation.

But despite these diagnostic challenges, all patients with arthritis of the
weight-bearing joints should be evaluated for limb-length discrepancies for a
number of reasons. A basic assessment is quick and easy, generally requires
no expensive tests, and can be done during a routine examination. Treatment can sometimes produce a dramatic and immediate reduction in symptoms...especially when the knees and hips are affected. And having a patient
temporarily try a lift in the shoe on the shorter side...while making sure that it
is less than the actual discrepancy found...is about as safe and easy a therapy
as any we have in medicine.

Once the examination has shown that a patient feels better with a lift during
testing, I begin by putting a temporary quarter-inch heel lift in the shoe on
the patient's shorter side, as long as the discerned difference is much greater
than this. It is most comfortable to place the lift under whatever insole is
already in the shoe. Although at times, not nearly fully corrective, even a
small difference can often bring some degree of comfort...which again, can
sometimes be felt immediately.

If a functional discrepancy is confirmed after seeing a physical therapist or
other specialist, specific exercises are recommended. In many cases these exercises will help eliminate the leg-length difference and hold the correction. I
have seen differences of more than an inch (2.54 cm) become fully corrected,
at which time a lift is obviously not needed. At other times a temporary lift
can still be used if it helps reduce symptoms during treatment, or if the treatment does not fully correct the problem or enable sustained correction.

Structural leg-length discrepancies are different. Although exercises will not help, permanent lifts usually do. But it is very important to understand that the body compensates for structural differences over a period of time. So even if exact differences are determined, corrections must be made very gradually to avoid additional problems.

After many years of functioning with unequal leg lengths, most people are not comfortable with full correction, even if those corrections were made in small increments. For this reason, my corrections are based more upon what I find to be clinically indicated and tolerated, than the specific amount of correction a radiograph may indicate. Long ago, I learned that as a physician, you should not treat an X-ray or blood test...but a patient. I still believe nothing beats clinical medicine.

In most instances of structural discrepancies, I begin by recommending a quarter-inch heel lift for about a week, just as I do with functional leg length differences. Again, this is done only if differences are found to be much greater than a quarter-inch. Although this very rarely occurs, the lift should be removed immediately if it causes any additional symptoms, such as low back pain. Then, if greater corrections are indicated and no additional discomfort has occurred, I have the patient try a quarter-inch lift inside the entire shoe. This can be done by using the insole from another shoe or purchasing one. Lifts up to about a quarter of an inch can often be tolerated in casual or running shoes. This is far more practical and less costly, and therefore certainly more preferable to putting a lift on the sole of the shoe. Extra depth shoes can be very helpful in this regard.

When additional corrections are required that can't fit inside a shoe, a lift on the sole of the shoe may be necessary. A quarter-inch lightweight material, which extends the entire length of the sole of the shoe on the shorter side, is recommended. The weight of the lift is important, as a heavy lift will make walking harder and more fatiguing. It may also contribute to tripping. A good craftsperson can often cut a thick sole in two and place the lift between the two halves, rather than applying the lift to the bottom. This is often better cosmetically, as well.

A full-length lift is preferable to just a heel lift, as it increases balance and stability. Using only a heel lift can also cause a functional shortening of the calf muscle and result in other problems. These include increased pressure on the ball of the foot. I also recommend that until a final determination is made, the lift be put only on one shoe which the patient will try for a while. Then, when corrective measurements are finalized, lifts can be added to other shoes. This results in significant cost savings and is far more convenient for patients.

In summary, assessing leg-length discrepancies can be challenging. Some patients have both structural and functional problems while others have functional discrepancies that change. Treatment, consisting of shoe modifications to raise the shortened leg, must be gradual as compensations have developed and aggressive corrections can increase symptoms. But because of the profound, destructive effect these problems have on the weight-bearing joints, the data that supports correction, the sometimes immediate and dramatic reduction in symptoms, cost-effectiveness and simplicity of the treatment... *all patients with arthritis of the weight-bearing joints should be evaluated and treated for limb-length discrepancies.*

"*I've had severe back and neck problems since my early thirties. Later, I developed hip pain. No one could provide a reason for this and I just accepted it as a fact of life. Finally, I could hardly walk at all and had a hip replacement because of severe arthritis. But I continued to have serious pain. I went to see Dr. Pack who found that my left leg was significantly shorter than my right. He corrected my structural alignment problems and now I can do anything I want with no pain at all!!!!! I haven't felt this good since my twenties, and I am now in my seventies! The compromises people accept as part of the aging process are unnecessary. I am a true believer. Growing older doesn't have to mean joint pain and decreased activity. I am the living proof and am so thankful to him every day.*"

Elaine Sutherland
Greensboro, GA

MUSCLES AND OSTEOARTHRITIS

University of Georgia All-American and former Tampa Bay tackle, Matt Stinchcomb, had an injury that limited movement of his ankle and prevented him from getting into his three-point stance. It was thought that the bone spurs in the front of his ankle (much like those in an arthritic joint) were the cause of his limited motion. Told that surgery was his only solution, Matt consulted me in hope of finding an alternative. Matt never had to have the recommended surgery. And in a phone conversation not long after my treatment, he told me that besides "no longer having any problems," he also noticed "improved balance and stability."

You may never have needed to block an opposing lineman or been injured like Matt, but you could have the very same problem and not realize it. Joint motion can be restricted for many reasons and one of the most common is tight muscles. These can be a primary cause, or occur secondarily, as the result of an injury or arthritic condition.

It is also important to note that tight muscles in one area can restrict and cause pain in a joint or muscle elsewhere. For example, tight muscles in your upper leg can cause knee pain as well as pain in your lower leg. So, whether you're a world-class athlete, or simply someone struggling with arthritis, know that *the flexibility of your muscles can be a major factor in joint mobility as well as musculoskeltal pain…in the localized area…as well as elsewhere.*

Muscles play an important part in all arthritic conditions. The joints are the "hinges" of our skeletal system, designed to allow a very specific plane of motion. But it is our muscles that make the motion possible.

Joints are designed to allow motion in a certain plane, and it is important for the corresponding muscles to create and utilize movement specifically in that plane to avoid joint damage. The elbow, for example, is a hinge joint. If that joint is not injured or arthritic and is used as it is anatomically designed, you should be able to open and close your arm easily with a free range of motion without discomfort for many years… again, *as long as that movement occurs*

---

[6] Most joints are designed to move. Some however, like those of the skull, are technically classified as joints but allow no motion.

within the range of motion that joint was designed for. But twist your arm all the way in or out, and the motion at that joint is more difficult, since it's not really supposed to be used that way.

Abnormal alignment causes stress on joints that subsequently damages them. That's why pitchers (as mentioned earlier) tennis players and others, develop problems in their elbow joint... because they are continually moving them in an abnormal plane. *This is a very important concept, because as you have seen, it is the underlying cause of arthritis of the weight bearing joints.*

In summary, both abnormal skeletal alignment and muscle imbalances can cause joint pain. Each may be primary or secondary. One affects the other, and both need to function normally to enable pain and stress free motion.

EXERCISE AND STRETCHING

Like any lever, muscles need to function around a stable, optimally aligned structure. For us, that's the skeletal system, whose stability begins first and foremost at our base with proper foot positioning. Once optimal structural alignment is achieved, a fine balance must be maintained wherein muscles are kept strong enough to move a joint, while staying sufficiently flexible to do so. But in almost all cases that balance isn't present, because some of our muscles are used far more than we stretch them.

Patients often confuse exercise with flexibility. Many don't realize for example, that simply going for a walk tightens calf muscles, consequently decreasing flexibility, and that a proper stretching routine is just as important as the exercise itself.

But most people don't stretch at all, and, if they do, it's usually improperly. Over time their muscles become shorter and tighter, restricting joint motion. Often, patients who think their joint stiffness is entirely due to their age and arthritis, simply have tight muscles. Because of the importance of this, I'll discuss proper stretching later in this chapter, in the section on treatment.

The more active you are, the more you need to stretch. Rarely does an Olympic track and field event occur without someone pulling or rupturing a muscle or tendon...and these are the greatest athletes in the world...who also have access to the best trainers and coaches. Although most of us do not have the same muscle mass they do, or incur the same kind of physical stresses, many of our injuries and problems are also due to muscle imbalances...a disproportionate amount of muscle tightening in relation to flexibility.

MUSCLE SPASM AND GUARDING

Failure to properly stretch our muscles is not the only reason they can cause limited joint motion. The body responds to arthritis and other painful conditions with muscle spasm and guarding, which in turn protectively restricts joint movement. Sometimes, however, that response can cause far greater symptoms than the initial problem. That's exactly what happened to Matt Stinchcomb. His injury may have initiated his symptoms, but it was his tight calf muscles that were the subsequent cause of his continued limited motion. Once that component was treated, he was fine.

In most cases of osteoarthritis of the weight-bearing joints, improper skeletal alignment causes joint pain, which in turn triggers the type of muscular reaction that Matt experienced, further limiting joint motion. This causes more joint stiffness. Muscles get even tighter and the disabling cycle continues. That's why properly aligning one's structure by doing such things as equalizing leg lengths, correcting pronation, and relaxing tightened muscles...by proper stretching and other means...is often far more effective in creating pain-free motion, than simply taking medications to alleviate symptoms or decrease inflammation.

It's always better to treat the cause of any medical problem than to continually try to alleviate the symptoms that result from it. Then, if medication is still necessary, less will be needed and its effectiveness increased.

Patients who've had knee or hip joints replaced have certainly experienced this. Poor alignment caused by the shortening or lengthening that occurred during surgery, combined with postsurgical muscle splinting, can initially make joint motion quite a challenge. That's why physical therapy is such an important part of the rehabilitation process.

But let's remember that poor alignment usually initiated the problem, and is often worse post surgically. *So if these patients were structurally aligned before beginning therapy, they would have much less discomfort and find it much easier to move their new joint and subsequently, walk.* In essence, they would rehabilitate much faster.

This concept is so important that it's worth reiterating. *Patients with osteoarthritis of their weight-bearing joints have pain for two basic reasons: poor alignment which is often greatly accentuated by joint replacement surgery, and muscle tightness, usually caused by a combination of improper stretching and the body's protective mechanisms of splinting and guarding.*

Tight calf muscles are one of the most important conditions related to arthritis of the weight-bearing joints because of the variety and severity of problems they cause. Yet, much like leg-length discrepancies, this almost universal, generally easily corrected condition remains for the most part overlooked and therefore, untreated in patients with arthritis.

TIGHT CALF MUSCLES (GASTROC-EQUINUS)
DEFINITION

The calf consists of two muscles: the gastrocnemius—subsequently referred to as the gastroc—which originates above the knee and has two heads or branches, and the soleus, which originates below the knee. Both muscles insert, or have their end attachments in the back of the heel or the calcaneus, by way of the Achilles tendon. Knowing where the muscles begin and end helps determine which ones are tight.

A minimum of ten degrees of ankle movement of the foot upward and back toward the front of the leg (dorsiflexion) is needed to function properly. When an inadequate amount of motion is available at the ankle joint, the word equinus is used…from the Latin word meaning "relating to horses" … because of the plantar-flexed or downward position the foot assumes, much like the hooves of a walking horse (See Figure 33). When only the gastroc is tight, the patient is said to have a gastroc equinus, whereas if both muscles are involved, the condition is referred to as a gastroc-soleus equinus. Distinguishing these is discussed below.

[7] We are discussing and differentiating two types of muscular equinus. A third type of equinus, osseous equinus, is a condition in which a bony block, such as bone spurs or a joint fusion has occurred, that permanently limits ankle motion. This type of equinus is not addressed here.

INCIDENCE AND CAUSES

Usually seen as a congenital problem, a gastroc-soleus equinus is much rarer and is often associated with problems that limit motion at the ankle, such as injuries and arthritis, or with neurological diseases like polio. This condition also occurs normally to some extent in people with a high arched or *cavus* foot. It often causes children to walk on their toes and generally requires surgical correction.

Tightening of just the gastroc is far more common and occurs in almost everyone to some extent, especially as we age. Even children who seem to be quite flexible elsewhere can develop this condition which at times can be severe. Interestingly, I often find the presence of severe gastroc equinus in a syndrome with gifted children, both with and without ADHD (attention-deficit hyperactivity disorder). These children may be quite flexible elsewhere and also have over active reflexes.

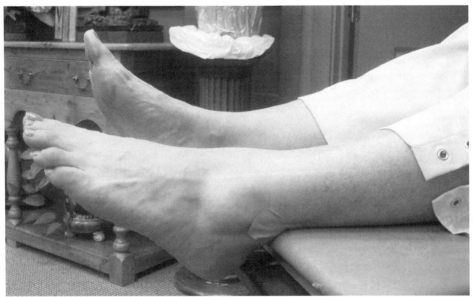

FIGURE 33
When the calves are not tight, it is easier for the foot to assume a position closer to 90 degrees to the leg. Tight calf muscles, however, pull the foot downward even at rest.

Notice the difference in position of this patient's normal right foot and the downward position of her left foot due to such tightening.

Positional problems can also cause tight calves. For example, women who wear high-heeled shoes often get a functional shortening of their calf muscles over time. This develops for the same reason that using a heel lift as opposed to a full foot lift occurs. The same is true of people who frequently drive long distances and have to hold one foot in a downward position for extended periods. Patients with a leg-length discrepancy can also develop tight calf muscles, usually more so on the shorter side. This is because they typically point their shortened foot downward, walking on the ball of the foot, as they try to more easily reach the ground. This eventually shortens their calf muscle.

Severely pronated or flattened feet can also functionally shorten calf muscles. Interestingly, the reverse is also true; i.e., tight calves are a major cause of pronation or flattening of the feet.

Excessive wear on the outside of the heels of walking or running shoes can cause overstretching of the calf muscles. A gastroc equinus can then develop as the body protectively splints and subsequently shortens a painful, over-stretched calf muscle. This was commonly seen years ago with the "Earth Shoe" craze, which resulted in many patients seeking the aid of their podiatrists.

Patients who are inactive or immobilized for a long time, like those wearing a cast, often develop this condition. On the other hand, extremely active individuals are also prone to developing an equinus because of the excessive tightening of these muscles. Since the calves tighten or contract with every step, even many relatively inactive people develop a gastroc equinus from not stretching properly. Failure to stretch, combined with the compensatory splinting mechanism that often accompanies pain, makes people with arthritis especially prone to developing this condition.

Surprisingly, another very common cause of tight calf muscles and joint stiffness is the Statins—the cholesterol-lowering medications millions of people take every day. Although these drugs, such as Lipitor and Lovastatin, are often effective, necessary, and even life saving, they also have serious and commonly overlooked side effects, especially in the senior population.

According to the Mayo Clinic [9], the most common of these is muscle pain. *Because so many people feel that aching and stiffness are part of getting older, they often simply dismiss these symptoms as arthritic, or even expect them as typically part of the aging process, never realizing the causal correlation with the drugs they're taking.* I notice this in patients very frequently..... far more frequently than reported. In the majority of these situations the calf muscles are specifically involved.

In many cases, despite usually very effective therapy, patients who take cholesterol-lowering medications can't keep their calf muscles properly stretched and symptoms often continue or even worsen…again, due to the side effects these drugs cause. Many patients, who have consulted with their physicians and have stopped taking these medications, have had much greater success not only in reducing their symptoms, but in feeling that their arthritis has been greatly improved.

This is not to imply that you should ever stop taking these or any prescription medications without first consulting your physician. To do so could have dire consequences. In no way, am I suggesting that ever be done. Used properly for the right indications, prescription drugs are irreplaceable. That said, if the option exists to avoid taking them, it may certainly be worth considering and discussing with your doctor.

DIAGNOSING TIGHT CALF MUSCLES

Diagnosing equinus is important, because among other things, patients compensate for this problem by pronating, and as you have seen, this is a major cause of osteoarthritis of the weight bearing joints.

Much data exists validating the casual relationship between equinus and musculoskeletal problems. R. S. Hill [10], in an article in the *Journal of the American Podiatric Medical Association*, states that "there is a significant correlation between compensation (pronation) for ankle equinus and podiatric pathology (foot problems)." He further adds that treatment of this condition can significantly reduce both the need for foot surgery and the complications that sometimes occur.

In a 2002 study published in the *Journal of Bone and Joint Surgery*, DiGiovanni and colleagues [11] also confirmed the direct correlation of equinus with foot pathology.

Further, Lavery, Armstong, and Boulton [12] reported that patients with equinus had a three times greater risk of elevated pressure on the bottom of the foot. While this has significant implications for everyone, it is especially important for diabetics, because reducing this pressure can greatly decrease the risk of foot ulcerations and amputations.

Barrett and Jarvis [13] reported an increased incidence of nerve entrapments in the foot with equinus, while a report by Szames et al., [14] in the journal *Clinical Podiatric Medicine and Surgery,* emphasized the significant impact of equinus on Sever's disease, an inflation of the growth center at the back of the heel in children. Meszaros and Caudell [15] reported that equinus played a significant role in the formation of both children's and adult's flatfoot deformities.

The effects of equinus are not limited to foot pathology. Higginson and co-workers [16], for example, reported that equinus can cause the knee to function in an abnormal position of hyperextension, while a Washington University study [17] showed that equinus can limit both knee and hip extension.

The diagnosis of tight calf muscles can be made in a number of ways. I'll explain how physicians and other professionals evaluate patients for this condition, and then I'll discuss self assessment for it, and show you an easy way to tell whether or not your own calves are tight.

First, it's important to determine if one or both calf muscles are involved because a gastroc-soleus equinus or tightening of both muscles often requires surgery, whereas, in the far more common case of a gastroc equinus, conservative treatment is often very effective.

Physicians generally examine patients for equinus by having them lie flat on their backs with their legs fully extended. Based on one line, imaginary or drawn, that bisects the lateral or outside of the leg, and another one along the outside of the foot, the angular position of the foot is noted. The closer that angle is to 90 degrees, the better. In patients with a severe equinus deformity the angle of the foot relative to the leg is greater than 90 degrees, meaning that the foot is in a more downward position, called plantar flexion. A simple visual examination on both sides often reveals the tighter calf, because the muscle contraction in the calf, tends to pull the foot into a more downward position. Once shown, patients can readily see this (See Figure 34).

Next, with the patient in the same position, and the leg being examined fully extended, and the foot held slightly supinated or turned inward, the examiner pushes the foot back toward the front of the leg without any help from the patient. A measurement is then taken. This exercise is repeated with the knee flexed or bent, and another measurement is made. The minimal amount of motion required for normal function is 10 degrees of dorsiflexion.

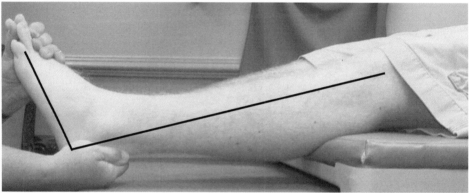

FIGURE 34A

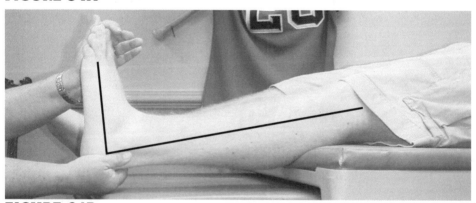

FIGURE 34B

Figure 34 shows the lines which are used as landmarks, described above. The top picture clearly shows an inadequate range of motion. While the one on the bottom shows improvement, with the foot reaching 90 degrees of dorsiflexion, another 10 degrees is actually necessary to function properly

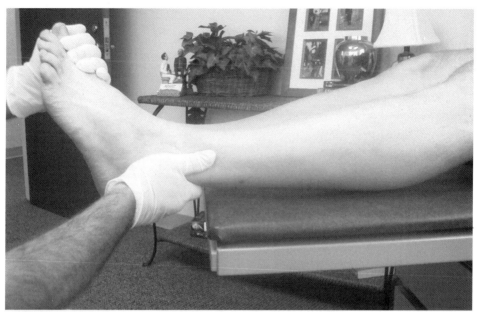

FIGURE 35A

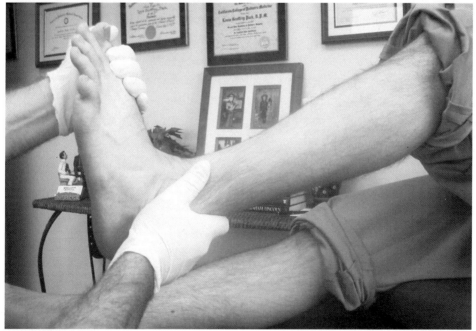

FIGURE 35B

Figure 35A shows the inability to flex the foot at the ankle with the knee extended, while Figure 35B shows an adequate amount of motion obtained by bending or flexing the knee, indicating a true gastroc-equinus.

This means that the foot must bend back towards the front of the leg, 10 degrees past a 90 degree position of the foot to the leg. If this amount of motion is unavailable with the knee flexed or extended, a gastro-soleus equinus is said to be present. With a true gastro-equinus, adequate dorsiflexion is lacking with the knee extended, but present when the knee is bent...again, a far more common and much more easily treated condition (see Figure 35A, and B).

SELF-ASSESSMENT

Do You Have Tight Calf Muscles?

Although you obviously can't perform a professional examination on yourself, there is a very simple way you can tell just how tight your calves are. Because a tightness of these muscles makes it difficult to flex the foot at the ankle, patients with this condition compensate by picking their toes up instead. This allows their foot to clear the ground and avoid tripping.

Walk barefoot and look down at your feet. With each step, see if you are actually flexing or bending your foot at the ankle, or just picking up your toes instead. Called "extensor substitution," this is one of the most dramatic changes that occur with an equinus, and is perhaps the easiest of all ways to identify this condition (See Figure 36A, and B).

Now take a look at your shoes. If you have an equinus condition, the ends of your shoes will sit up, off the ground (See Figure 37A). This is due to the repeated lifting of your foot by your toes because motion is unavailable at your ankle. You can sometimes place a number of fingers under the end of the shoes where the toes are. Many try unsuccessfully to rely on shoe trees to prevent their shoes from rolling up in this manner, unaware of the true reason this is developing.

There may also be a prominence, or even a hole, in the tops of the shoes from the constant pressure at the area of your big toe as it forcibly tries to lift your foot with each step (See Figure 37B). People with this condition often like to wear sandals so their toes are free to move up and down.

Patients with tight calves are also frequently much more comfortable wearing shoes with a heel lift. This decreases the pull on those muscles that is often accentuated by wearing flat shoes, and makes them much more comfortable. But in so doing, a greater degree of functional shortening occurs.

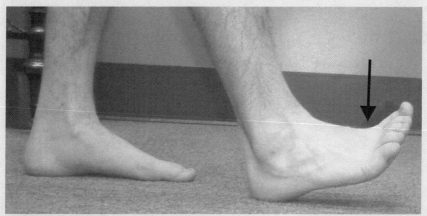

FIGURE 36A

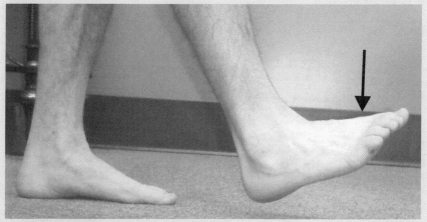

FIGURE 36B

The patient in Figure 36A is unable to flex or bend his foot at the ankle due to a gastroc-equinus. Instead, with each step he brings his toes up to avoid tripping. Figure 36B shows the same patient after proper stretching. He is now able to walk normally, by bending his foot up at the ankle, rather than having to lift his toes.

SELF-ASSESSMENT
Do You Have Tight Calf Muscles? (cont'd)

FIGURE 37A

FIGURE 37A
These are the shoes of a patient with a gastroc-equinus. Notice the end of the shoes sitting up, off the ground.
FIGURE 37B
The typical location of a hole in the top of a shoe of an individual with an equinus deformity.
FIGURE 37C
With severe pronation and equinus, the distal medial aspect of the soles of the shoes shows the greatest wear.

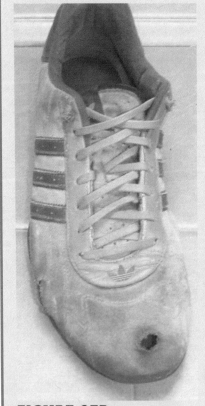

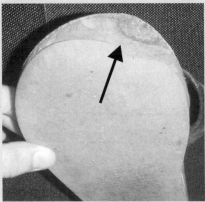

FIGURE 37B

FIGURE 37C

GAIT ANALYSIS

Gait analysis, the study of a person's walk, is very revealing in patients with an equinus, and many other conditions. Tight calf muscles cause an "early heel-off"…that is, because of the shortened calf muscle, the heel is lifted quickly and prematurely. The foot may also seem to turn outward with each step. This is called an "abductory twist." Excessive pronation or flattening of the foot is also common, as is reduced gait velocity, and other, more technical changes.

TRIPPING AND FALLING

Because lifting one's foot is difficult with tight calf muscles, the shoes of these individuals will also show excessive wear at the very tips of the soles (See Figure 37C). This indicates that with each stride they are basically hitting or kicking the ground with the ends of their shoes. Because of this, patients often feel clumsy, their feet tending to get "stuck" when they walk. Many trip and fall as a consequence. This is particularly true when walking on carpet… especially with thick, ridged, rubber-soled shoes.

Although often associated with a neurological disorder, or a loss of balance as part of our natural aging process, tripping and falling are sometimes due to nothing more than tight calf muscles. This has widespread implications because falls are the leading cause of death by injury in older adults. Only half of such hospitalized cases are alive after 1 year. It has been estimated that by 2030 the number of Americans that will die annually from falls will reach 280,000 [18]. So be sure to stretch your calves!

OTHER SYMPTOMS

Besides tightness or soreness in the calves, equinus can cause similar symptoms in the fronts of the legs, much like the shin splints that are seen in runners. That's because the muscles in the front of the leg must work much harder…against the resistance of any tightened calf muscles…when attempting to lift the foot while walking or running. Pain in the front of the ankle joint can also develop because of the jamming effect created for the same reasons.

As you've seen earlier, equinus also causes a great deal of strain on the Achilles tendon which attaches the calf muscles to the back of the heel of the foot. As the pull on this tendon increases, patients may develop a number of problems, including bursitis and insertional tendonitis, sometimes referred to as Achilles tendinopathy. With continued stress, large bone spurs can form in the back of the heel. These can be very painful and disabling, at times requiring surgical correction. If enough stress is placed on this tendon, it can rupture, completely tearing away from its bony attachment.

As mentioned, the major effect of equinus on the foot is pronation, so all of the problems associated with the later may be seen when the former is present. Many of these conditions have been discussed and include, but are certainly not limited to bunions, plantar fasciitis (pain under the arch), heel spurs, neuromas (enlarged nerves between the toes), tailor's bunions (dislocations seen as bumps on the outside of the foot behind the little toes), and many other problems.

TREATMENT

There are many therapeutic modalities that can be used to remedy tight calf muscles. As mentioned previously, since this is not a medical text, I will only focus on proper stretching techniques which are generally quite effective. Even individuals with a true tightening of both calf muscles…again, a much rarer condition…may benefit from stretching their calves as described in this section.

A tightened muscle needs to be stretched, and there are many ways to do this. Remember, if tight calf muscles were not as prevalent and troublesome as they are, there wouldn't be as many products on the market for stretching them…slant boards, special shoes, night splints, and all sorts of other things. The calf stretch that I have found to be most effective was actually shown to me by my son Jeff, and can simply be done on a step.

1. STEP STRETCH

A muscle stretches best if it is alternately stretched and contracted. Begin by standing on a step while holding onto a hand rail. It will be more comfortable to do this when wearing your shoes and foot inserts, if you have them.

This patient is demonstrating the correct foot position for stretching the calf muscles on your right side.

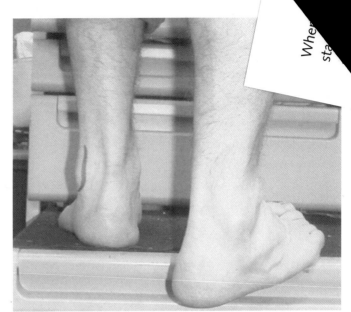

FIGURE 38A

Whe... sta...

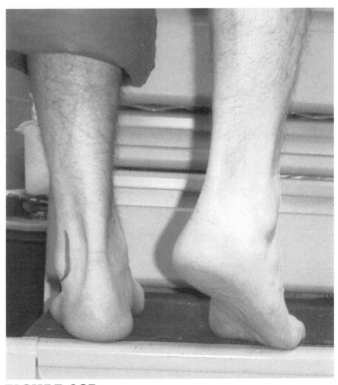

FIGURE 38B

stretching your right foot, place your left foot totally on the step, as if riding normally. Move your right foot back so that only the ball, or about a third of your foot, is on the step.

Traditional stretches are done by dropping the heel with the foot in this position. But in people who pronate excessively, the foot will roll inward usually causing this stretch to be somewhat ineffective. For maximum effectiveness, begin by standing with your right foot turned out and away from your body about 45 degrees, while it remains flat on the step. If your foot still has a tendency to roll inward, the placement of a small wedge, like a doorstop, under the inside of the ball of the foot, will help keep the foot in its correct position. You may even want to try stretching both ways, so you can really see and feel the difference this recommended position makes. The key is to not let your foot roll inward or pronate as you stretch.

The knee on the leg you are stretching or contracting should always be straight, while the other can be bent or flexed for comfort. For example, while attempting to stretch your right calf, your left knee should be bent and your right knee kept stiff, to increase the stretch. Then let your heel *slowly* drop downward to the floor (See Fig 38A).

If you haven't done this in a while, and there is significant tightness, dropping the heel too rapidly can cause discomfort and injury. Begin by trying to push the heel down to the floor while applying more and more of your body weight. The key is to keep a slow, gradual stretch and to not bounce. If pain occurs, back off and proceed more gradually in smaller increments. Remember...pain causes muscle spasm and guarding, which is what you want to eliminate. It may have taken many years for you to become this tight, so it is ok if it takes a few weeks to really get these muscles stretched properly.

Stretch and hold this position for about 20 seconds. Then stand up on the ball of your right foot so that your calf contracts or tightens (See Figure 38B.

This only needs to be done for a much shorter period of time (5 seconds or so). Do not turn or lean on the outside of your foot when doing this. Just go straight up on the ball of the foot while placing more weight on the side you are working on. You also do not need to go up very high when contracting your calf…just enough to tighten it. If you go up too high on your right foot, you will also have a tendency to go up on the ball of your left foot. This will tighten the left calf instead of relaxing it. Remember, the side you are working on always stays straight; i.e., the knee on that side is not bent at all, while the other leg is bent at the knee. This will maximize the stretch and contraction.

This cycle is considered one set. Complete five sets while alternately stretching and contracting your calf. Then do the other foot the same way. Optimal flexibility can be achieved by completing five sets of these exercises, which is about 2 minutes on each side, two or three times a day. This stretch should also be done before, after, and *during* any break in exercise.

If you muscles are really tight, initially taking a short break while exercising will really be very beneficial too. For example, if you are going for a walk, stop and stretch periodically. You do not have to do five full sets on each side. Even a couple of sets will help you stay more flexible. It will also prevent excessive tightness that would otherwise occur at the end of your activity, and be much more difficult to correct. Note that these times and amounts to stretch are all simple guidelines. Whether you stretch 18 or 22 seconds will not change your results significantly.

Once your calf muscles have been properly stretched, you will not have to do this as often. Stretching your calves, as well as your other muscles, should always be done every day. Remember, you can't race a car without changing the oil.

2. SITTING STRETCH

If a step is not available, or standing on one too difficult, or causes too much pain in your calf or the bottom of your foot…as sometimes occurs with heel spurs…then there is another way you can stretch. Although not quite as effective, this will still be very helpful.

Sit flat on the floor with your feet and legs straight out and your back against the wall. Turn your right foot slightly inward and place a belt around the ball of your foot. Note that with your foot in this position, your left hand would

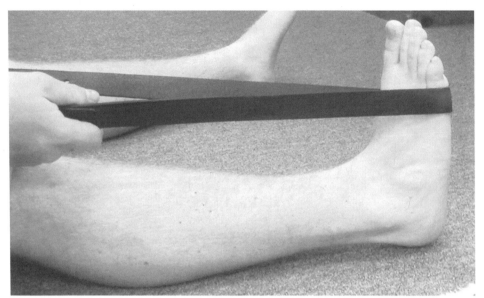

FIGURE 39A

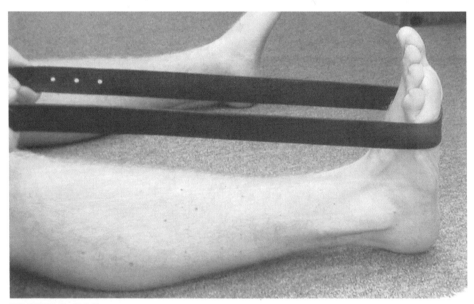

FIGURE 39B

This patient in figure 39A, is demonstrating the proper way to stretch calve muscles while sitting on the floor with his foot turned slightly inward so as to not pronate it, while Figure 39B shows the foot in an improper, pronated position

be a little closer to you than your right. Now, keeping your foot and leg flat on the ground and using *only your arms,* hold the ends of the belt and pull, bringing your foot back toward you by letting it bend at the ankle. This is the stretch phase. Hold this position for about 20 seconds and then using *only the power of your foot and not your arms,* push against the belt for another five seconds (See Figure 39A, and B). This will contract your calf muscles. Repeat this cycle five times (five sets), and then do the other side. Remember to try and keep your back straight while doing this and to not lean forward. This will give you a better stretch and at the same time prevent you from hurting your back.

3. LEANING STRETCH

If you are exercising away from home where there is no step available, you can also stretch your calf muscles by facing a wall, and leaning into it with outstretched hands. Place your right leg further back and make sure your right knee is not bent at all. Now bend your left knee (See Figures 40A, and B).

For this exercise to be most effective, remember to lean on the outside of your right foot, making sure it is not flattened or rolled inward. Hold this position for approximately 20 seconds. After completing this step, go up on the ball of the right foot for five seconds. This will contract the calf muscles. Be sure to keep your left foot flat on the ground. Now repeat the process on the left leg. For the best result, complete five sets on each leg.

Although each of these three stretches is helpful, the Step Stretch is much preferred. Because all muscles stretch better when they are not cold, doing these exercises after a hot bath or shower will make them much easier.

In cases where the calf muscles are strained or injured or where there is simply just too much shortening, try wearing small heel lifts in your shoes on the affected side or sides. This can reduce the pull on the muscles by functionally lengthening them. Generally, this should not be used on an ongoing basis.

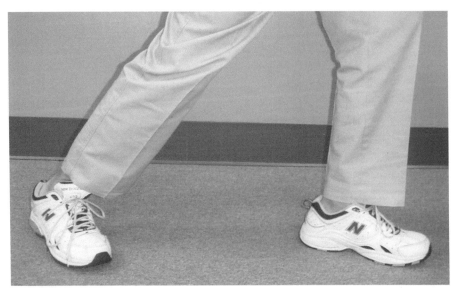

FIGURE 40A

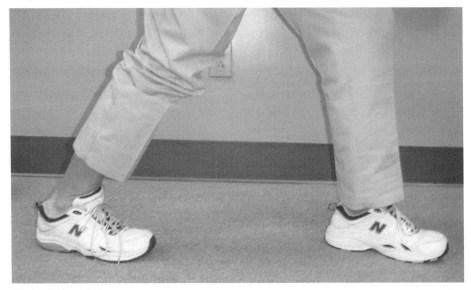

FIGURE 40A

Figures 40, A and B show a patient stretching their right calf while leaning against a wall. Many people do this stretch improperly as depicted in Figure 40A. Notice how the patient's right foot is pronated. This will not allow the calf to stretch fully. Figure 40B illustrates the proper, slightly supinated position of your foot while doing this stretch. Remember, you should be leaning on the outside edge of that foot as shown here.

"*My wife could not walk a half mile without severe foot pain and no longer enjoyed golf or tennis, so she scheduled surgery with a prominent Atlanta doctor. Hearing of Dr. Pack, we cancelled the surgery. He diagnosed structural problems. A few days later we went to France for a long vacation. Because of her serious problems, I had arranged for a car and driver and even a wheel chair. With her specially designed foot orthotics that he made, which arrived the day before we departed, she walked several miles on Paris sidewalks – with no pain! Yesterday, she played in a golf tournament – no pain. And besides pain relief, her performance has improved! Her swing is stronger as she is no longer shying away from shifting her weight. So I decided to have him treat me too. I am hitting the golf ball longer as my stance is now level... perhaps for the first time in my life. For a couple of active retirees, eliminating pain and improving our sports life is a big deal! I wish we had known about him years ago.*"

Dennis Berry

Former President and COO, Cox Enterprises and former President and Publisher, Atlanta Journal-Constitution

Now let's see how to really fix pronation and stop arthritis!

Tim Mack brings back thanks, wearing his Olympic gold medal and holding the special custom foot orthotics that Dr. Pack designed for him.

CHAPTER 7

The Fix

A BOTTOM up APPROACH

"After winning the Gold Medal in Athens, the special custom inserts Dr. Pack made for me mysteriously disappeared. I wanted to set a new world record so I had him make me another set."

Tim Mack

1995 NCAA Indoor Champion, 2001 Goodwill Games gold medalist, 2002 U.S. indoor champion, 2004 Olympic trials champion, 2004 Olympic gold medalist, 2010 U.S. Indoor Champion (age 37 years old), pole vault.

- Your ultimate success in eliminating pain, staying active, and preventing further joint damage, first depends on your acceptance of the premise that osteoarthritis is caused by structural malalignment.
- Although there is no known cure for osteoarthritis, there is no better, faster, more effective way to prevent or stop its progression, than the use of precisely made prescription foot orthotics. They counter act the mechanical factors that are the root cause of arthritis of the weight-bearing joints.
- Despite the above, corrective, custom made, foot orthotics are not fully utilized in the treatment of arthritic conditions.
- To truly alter function, a foot orthotic must be accurately measured and made. In most cases, the greater the degree of correction, the greater the improvement in function and reduction of symptoms.
- Making a fully custom orthotic optimally effective involves a thorough structural examination, a very accurate foot impression, corrective orthotic assessment, and follow-up evaluations.

You have learned a great deal about the real causes of osteoarthritis of the weight bearing joints, and have come to understand the important role the foot plays in this chain of events. Now you will see how effective something as basic as a custom foot insert can be... when done correctly.

If you're a senior citizen, you've walked the equivalent of four times around the world [1] (Kettering, n.d.) and placed more than 900,000 billion pounds of pressure on your feet (see "Foot Facts" in Chapter 4). Like most people, you've probably done nothing to optimally align yourself on a more permanent basis than periodic adjustments can provide, often worn the wrong shoes, engaged in physically stressful activities, not rested completely when injured, and been remiss in stretching. Are you really surprised that you're now paying the price with painful joints? Just imagine what the tires on your car would be like with that type of care. *Fortunately, for most of you it's not too late...there's still a lot that can be done to make you feel better.*

THE GOALS OF TREATMENT

Pain medication won't help much if you continue to bang you head against the wall. You must eliminate the root cause of any problem to have continued results. Unfortunately, most osteoarthritis treatment focuses on eliminating symptoms rather than the cause, and that's exactly why this disease continues to progress and affect more and more people.

Treatment may include methods such as shoe lifts to equalize leg-length discrepancies, physical therapy, massage, injection therapy, and surgery. But one of the most powerful and effective treatments is a precisely made set of foot orthotics.

Now before you jump to any conclusions and think, "but I've had *those*, and *they* didn't work for me," "I already have a set," or "How can just a pair of foot inserts really help the arthritis in my knees,?" know that I hear you and these statements every day, but please be sure to read on.

"This guy is a genius; a mad scientist kind of guy. What he says is so different but makes so much sense, you just have to listen to him. We're sure are glad we did! As soon my players put the special golf inserts - their "LouPack's" - in their shoes that Dr. Pack designed for them, they started winning and never stopped!"

Chris Haack,
University of Georgia Golf Coach

1999, 2005 National Coach of the Year, four-time Southeastern
Conference Coach of the Year, two National Championships,
seven Southeastern Conference Championships, 47 All-Americans,
51 All-Southeastern Conference Selections, 44 Team Championships,
31 Individual Medalists, seven Faculty Athletic Representative Awards,
six Team and Individual School Records , 2001, 2006 U.S. Palmer Cup Coach

IMMEDIATE GOALS

Some of the immediate goals of treatment include:

1. ACCEPT THE PREMISE

The first goal of any treatment is to clearly identify the problem, then determine why it's there (its cause), and decide what you are going to do about it (treatment). As you well know by now, osteoarthritis of the weight-bearing joints is a result of the abnormal frictional forces caused by continuous malalignment...more specifically, excessive rolling in and flattening of the foot at the subtalar joint. This is our old nemesis "excessive pronation."

And as I have mentioned...*improving your situation specifically depends on the acceptance of this premise as the root cause of your problems, and the redirection of your focus on achieving optimal foot alignment.* Remember, the more accurately you align the tires on your car, the longer they'll last and the better the ride you'll get.

Other modes of therapy may be helpful to some degree, *but nothing else I know of can so quickly, safely, and effectively reduce your symptoms, increase your function, sustain long-term results...and thus prevent or stop this disease...than an optimally corrective set of foot orthotics* (See Figures 41A, and B).

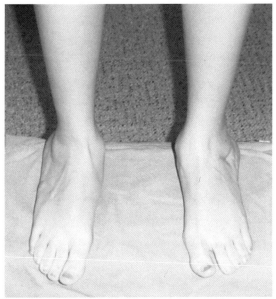

FIGURE 41A

This young girl has had juvenile rheumatoid arthritis, a true inflammatory disease, for many years. Although this particular type of arthritis is not initially caused by malalignment, the destructive nature of this disease has certainly resulted in it. Figure 41A on the left, shows the failed results of major surgery that attempted to correct her structural foot problems...which were causing serious knee, hip and back pain. Figure 41B below, shows her optimally aligned with fully corrective orthotics. As I said, nothing else can offer such a quick, dramatic, safe and effective result...even in a severe case as shown here. This patient went from being unable to actively play with her friends to now, as an adult, being able to function far more effectively. Certainly, her medications, physical therapy and other modes of treatment have helped her function better. But only the correction you see her standing on, finally made this possible. Her testimonial is in chapter 4.

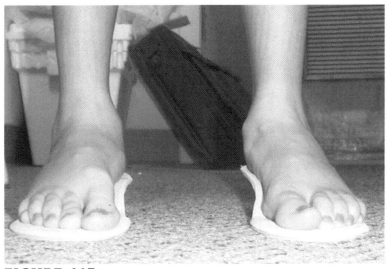

FIGURE 41B

As you have seen, there are currently two schools of thought regarding osteoarthritis: the generally accepted and long standing belief, that osteoarthritis is a disease of unknown causes...primarily associated with age and weight...and the much more recent and far more accurate concept, that abnormal alignment is a major factor in the cause and progression of it. And although the validity of the latter has been demonstrated in study after study and supports the principles presented in this book, it has not yet been fully accepted by the medical community.

But the answer is not rocket science. The remedy for osteoarthritis is very basic, powerful, sometimes dramatic, and can be administered by other medical professionals just as I have been doing every day in my office for a good many years.

2. FIND THE REAL CAUSES

Primary focus must be placed on finding and correcting the specific structural abnormalities that are the real causes of your disease...such as a leg-length discrepancy and excessive pronation... rather than deciding which medications or surgical procedures may be indicated because of the damage these problems have already caused.

This means having a structural evaluation by a professional who specializes in these types of problems. Unfortunately, because of the lack of importance generally attributed to these issues, getting a really qualified opinion isn't always easy. Podiatrists are experts that specialize in evaluating excessive pronation and other structural abnormalities. Certainly there are other qualified practitioners as well.

Any temporary measures that can be utilized to decrease or correct the causes of malalignment should be employed from the very onset of therapy, even while final determinations are being made.

3. ELIMINATE COMPENSATORY REACTIONS

As I've mentioned, one of the body's responses to pain, whether due to an injury or to arthritis, is muscle spasm and guarding. Although necessary and designed to prevent further problems, this protective reaction can cause far greater pain and restricted motion than the arthritic condition itself.

Physical therapy, therapeutic massage, hydrotherapy, and other modalities can be very helpful in eliminating such symptoms...*but as I've said...only if the underlying structural problems are being corrected at the same time.* It's a lot better to plug the hole in your boat first, rather than trying to continually bail it out. This is a very important, yet generally overlooked aspect of care in the treatment of those with arthritis of their weight-bearing joints.

That's why specialists like physical therapists and chiropractors usually see patients over such long periods of time...not because what they're doing isn't helpful or doesn't work, but because as I've emphasized, like drugs and surgery, treatment often can't be optimally effective or permanent unless the underlying structural causes are eliminated.

4. LIMIT WEIGHT-BEARING ACTIVITY...INITIALLY

Initially is the pivotal word here. Experts agree that one of the worst things you can do to a car is let it sit unused for long periods of time. But that doesn't mean you should drive it without (among other things) properly in-flated, balanced, and aligned tires. And the same is true for us. We should be properly aligned from the ground up beginning with our feet, *before doing* weight-bearing activities.

Authoritative sources, such as the American Rheumatology Association, agree that moving your joints is critically important to keeping them healthy. But what you're not often told is that *you must only use your joints in proper alignment.* Otherwise, the resulting increased friction will cause even fur-ther damage, far outweighing any beneficial effects. That's why as I've said earlier, "Health is something you can go through on the way to fitness."

Two main aspects must be treated: (1) The structural abnormalities that caused your joint problems initially, and (2), the painful symptoms that have resulted from your malalignment. Much like a sports injury, there's the injury itself that must be treated, and the body's response to it.

An injury requires rest to let the body heal. Returning to activity too soon after an injury is an all-too-frequent mistake made by coaches and athletes alike. This prolongs symptoms and risks permanent damage. As noted earlier, runners for example, are notorious for taking pride in the fact that they have "run through their pain," which only adds insult to injury.

So until you are optimally aligned, limit your weight-bearing activity. Then you should be able to function far more effectively with less discomfort. In the meantime, swimming and other non weight-bearing exercises are preferable.

5. EVALUATE MEDICATIONS

If symptoms are very severe or prolonged, oral medications may be indicated. Again, I'm certainly not opposed to using pain or anti-inflammatory drugs when necessary...they can be very helpful. Periodically, injection therapy can also be quite effective. Indeed, I have been impressed many times with the immediate reduction in symptoms in an inflamed joint or tendon sheath following a localized cortisone injection.

Two other types of injections are also safe and effective. For various inflammatory conditions of the foot, symptoms can be dramatically reduced by using a posterior tibial periarterial block, [2] which is an injection of a local anesthetic on the medial or inner side of the ankle. Introduced to me by my mentor, Dr. Marvin D. Steinberg, I certainly cannot take credit for this wonderful therapeutic measure.

This particular injection causes a profound increase in blood flow to the foot and can be used in a wide variety of inflammatory conditions, such as heel spur syndromes, metatarsalgia or pain in the balls of the feet, gout, and even fractures, to increase healing.

Another helpful type of injection involves a trigger point that I believe, lies just behind the ankle on its lateral side. In cases of severely tight calf muscles, injecting a small amount of local anesthesia into this area can help relax these muscles so they can be properly stretched. Although the anesthetic is certainly temporary, breaking a cycle of spasm can help long term when proper stretching is continued.

As stated earlier, when medications and physical therapy are used at the same time that causal factors are being eliminated, the synergistic effect can be quite powerful...that is...all therapies will be needed less, and each will have a far greater effect. This will also result in lower costs and reduced risks of complications.

LONG-TERM GOALS

Once underlying alignment problems are being treated and symptoms begin to decrease, there are a number of things you can do to function better long term. Some of the long term goals of treatment include:

1. STAY ACTIVE

Walking and other weight bearing exercises are preferable...as long as they don't cause symptoms. At times, even after optimal alignment is achieved, exercises with less joint stress, like water aerobics, may be more appropriately employed and or continued instead of those that are weight-bearing.

2. MAINTAIN OPTIMAL WEIGHT

As you have previously seen, excessive weight is not usually a primary cause of joint pain. It is, however, an important secondary factor. So if you want to continue to decrease your symptoms, it is very important that you keep your weight in check even after you have been optimally aligned.

3. DEVELOP STRENGTH AND FLEXIBILITY

I have noted that good muscle tone and flexibility are necessary for proper joint function, especially for those with arthritis. Unfortunately, most of us have muscle imbalances. Proper strengthening and stretching exercises help correct these imbalances, better align our joints, and thus decrease destructive frictional forces.

Yet many patients neglect this aspect of their care. Often this happens because they either don't realize its importance, or don't know how to do it without hurting themselves. For example, most people who stretch, do so improperly, and don't realize it. So it may be advisable to get the advice of someone like a trainer or physical therapist...not necessarily on an ongoing basis…but at least initially. With a little work, improvement in these areas can usually be accomplished at any age and make quite a difference in how you feel and function.

4. ELIMINATE MEDICATIONS

By treating the underlying mechanical causes of your arthritis…perhaps for the first time…the long term goal is to eliminate as much medication as you can. Obviously, this will not be possible or preferred in all instances, especially in severe cases, but you may be pleasantly surprised as to how much less pain medication you need once you are properly aligned.

ORTHOTICS

The Leaning Tower of Pisa didn't get crooked from the top, but from a poorly laid foundation. So fixing it isn't best done by splinting or cutting out sections of concrete from its middle. Like the tower, it's far better to correct our alignment problems at our foundation, rather than attempting to do so with knee braces and surgical procedures. And the easiest, most effective way to do that is with fully custom, optimally controlling, foot orthotics.

Orthotic is a very broad, generic term, much like the word *car.* Although simply defined as "a vehicle moving on wheels," the later could actually refer to anything from a three-wheeled econobox to a Ferrari. So I have never ceased to be amazed when patients tell me that they *"have already had inserts,"* and *"they didn't work."* All this really means is that these patients were putting something in their shoes. Exactly *what that something was*, its specifications, and how precisely it was constructed, makes all the difference in the world.

So if you've had one or more disappointing experiences with some sort of foot device, …please… don't assume that *"they"* don't work…anymore than you would assume that cars don't work because you once had a lemon. I say this fully aware of the fact that many of you have had countless bad experiences. Indeed, patients often come to me with shopping bags full of totally ineffective foot devices they have wasted their money on…and I'll explain just why this happens later in this chapter.

Because most people don't understand why one orthotic is better than another, their purchasing decisions are frequently based entirely on price and insurance coverage. While these are certainly important considerations for all of us...especially in today's economic environment...that's a big mistake. Although reality can indeed be disheartening, *the fact that you're insurance may cover your foot inserts, doesn't necessarily mean you're getting anything of real value for your arthritis.* A coupon for a free meal certainly doesn't indicate the quality of that meal...you may even get sick from it!

When you purchase a foot insert, you are not just buying a product, but an arthritis (and or sports performance) treatment. While orthotics are only one of the means to an end, they can be the most important and fastest acting. The ultimate success you achieve with them depends on many factors that are incorporated into this type of product.

Unfortunately, there is a great deal of confusion and even more misrepresentation regarding the various *"things"* that can support the human arch. Among others, numerous descriptive terms such as orthotics, foot braces, arch supports, insoles, devices, shoe inserts, pads, lifts, molds, and inlays are used. So many, in fact that those using them, often don't know what they have or what to call them. Indeed, even physicians can become confused.

Foot devices are available through many different venues from over-the-counter store-bought items and mail-order products to the fully customized devices obtained from physicians' offices. The latter can involve varying degrees of correction and flexibility, and because of this, additional terms such as *functionally corrective* and *semi-rigid* may be used too.

Since most people have some flattening of their feet when they stand, almost everyone could benefit from something in their shoes to support their arches. Because of this, and the fact that generic foot devices can often be made *to look like fully custom ones,* patients...particularly arthritis sufferers...are easy prey for the opportunists who know this all too well (See Figures 42A, and B). Unfortunately, the marketplace is inundated with products that promise miraculous results, but deliver nothing of the sort.....truly reminiscent of the old elixirs sold in the Wild West.

[2] As you have seen from the testimonial above, I designed special golf inserts for the UGA Golf Team the year they won the national championship (2005). Like many people, the players didn't know what to call them, so they combined my first and last names, often asking one another if they were wearing their "LouPack's"

Although certainly not always the most authoritative source, but rather a compilation of public knowledge created by those with access to the internet, Wikipedia [3], defines an *arch support* as, "an appliance designed to provide underside support for the foot." A more correct term for something that supports the arch is *orthosis*, the plural of which is *orthoses*...a Greek word meaning "straightening." But frankly, it has become customary over the years for most professionals like me, to use the technically incorrect term *orthotic*, when referring to such devices. It should be noted that there are many other types of orthotics, such as those used in dentistry. Obviously, I refer here only to those used in shoes.

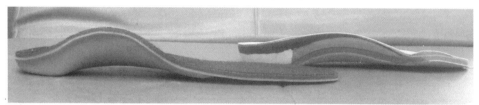

FIGURE 42A

Various types of foot inserts may look quite similar, but have totally different levels of effectiveness. As a matter of fact, the fit and finish of the mass produced, prefabricated, nearly ineffective insert seen above, is actually much nicer looking than the far more controlling, handmade, fully corrective one below.

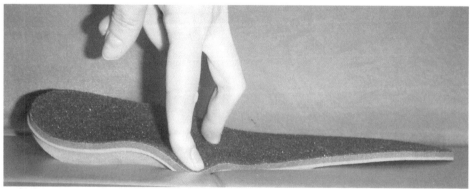

FIGURE 42B

Think about it...how useful is a device that will collapse with one finger's pressure? Comfortable perhaps...but structurally corrective... not possible.

From the time we first began wearing shoes, different types of products and substances both fixed and removable have been used to add comfort. Early innkeepers for example, attempted to alleviate the discomfort travelers experienced from wearing poorly cushioned shoes by making insoles of matted animal hair.

The real breakthrough in today's orthotics began in the late 1960s and early 1970s, when podiatrists routinely began to study biomechanics and perform gait analysis. At that time thermoplastics had been developed that could be heated and then cooled to make a thin, strong, very durable insert to control abnormal function. I remember well, making these now seemingly prehistoric devices that we then thought were state of the art... and at that time, they were.

Today, the orthotics I design and use still begin with taking a very accurate impression of each foot in its optimal functioning position and then having every single device custom-made to exacting specifications.

I continue to be impressed with the effect that even minimal amounts of orthotic correction can have on joint pain. Sometimes, even in very difficult cases, where I've made inserts for patients who are still having problems, adding only a few degrees of correction can cause dramatic improvements.

This occurred with the patient whose testimonial appears below. A very active 74 year old male, he has what appears on X-rays to be nearly complete obliteration of his big toe joints (See Figures 22, 23A,B,C,D, and E on pages 73 & 74, for a complete description. Figure 23D is repeated on page 134). The limited motion, severe pain and X-ray findings would certainly indicate the need for total joint replacement with no other alternative. Indeed, this is a text book indication for such treatment. Although not possible in all cases, he had a truly amazing result with nothing more than some adjustments to the inserts I had made him and some basic stretching exercises.

"Dr. Lou Pack is a brilliant diagnostician whose special insight into the relationship between arthritis and physiological structure has significantly improved an important aspect of my health. I had severe, sometimes disabling pain due to the arthritis in my big toe joints and feared the need for joint replacement surgery. I had almost no joint motion and what little I did have caused extreme pain. It significantly limited my ability to do many of the things I enjoy, including gardening and walking. His analysis of the true mechanical sources of my foot problems, the custom foot orthotics and specific stretching exercises he prescribed, have totally eliminated the pain in my feet or any other joints. I now have a much greater range of motion in these joints and can once again do normal physical activity without experiencing any discomfort whatsoever...absolutely none! Dr. Pack's pioneering work in this field should be a trailblazer to revolutionize the treatment of arthritic disorders."

Fred Kaplan

Distinguished Professor of English Literature at Queens College and the Graduate Center of the City University of New York, author of the critically acclaimed biographies Gore Vidal, Henry James, Dickens, and Thomas Carlyle, which was nominated for a National Book Critics Circle Award and a Pulitzer Prize. He has held Guggenheim and National Endowment for the Arts Fellowships, and was a Fellow of the National Humanities Center.

TYPES OF ORTHOTICS

There are two general categories of foot inserts: soft and hard. These classifications are broad, since soft devices can be reinforced to add much more stability, and more rigid devices can be significantly cushioned.

Soft, more flexible devices are used simply to pad the arch and provide some comfort, although at times they can help relieve painful foot symptoms. Harder, more rigid devices alter function, and can be further categorized as semi-rigid or fully rigid. Each can include additional padding that can be placed around bony prominences...such as on the ball of the foot...to alleviate symptoms in arthritis sufferers. Both soft and hard devices may be prefabricated or completely custom made.

To decrease arthritic symptoms, especially in the weight-bearing joints above the foot, abnormal function must be altered, and this can only be accomplished with devices that have some degree of rigidity, and angulated correction. But rest assured...this certainly doesn't mean that they can't be comfortable, as well.

By definition, a true orthotic or orthosis should straighten or hold you in a more upright position and correct your abnormalities...not merely cushion the foot (See Figures 43A, B, and C). Although almost all foot products provide some degree of comfort, very few *actually provide true structural alignment.* Those that attempt to do so are usually not optimally corrective.

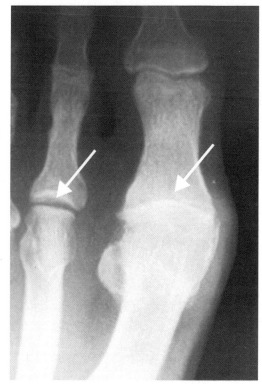

FIGURE 23D

Complete bony obliteration of the big toe joint space, with bone spur formation, is seen in Figure 23D. Notice the normal space in the joint directly next to it.

These are the reasons that painful symptoms are often unaffected or not more fully improved.

Prefabricated devices of any type cannot be maximally corrective, because they are not made for your particular foot structure and problems. Remember too, your feet are not identical, and therefore function differently to some extent. So a prefabricated set of devices can't possibly work exactly the same for both your feet.

One of the hallmarks of a good functioning orthotic—the key—is to be able to hold the ideal correction that was captured while sitting, when the patient stands and bears full weight. Unfortunately, most patients who wear foot inserts are far from optimally corrected. Many stand just as flat footed on their devices as they are off of them. Remember that optimal foot correction is directly related to decreased stress and improved symptoms in all the weight bearing joints the foot supports…and that's exactly why this is so critically important.

THE BENEFITS OF ORTHOTICS

We know that foot inserts have value. Shoe manufacturers have learned that adding some sort of soft padding under the arch to help support the foot really adds comfort and helps them sell their products. That's why athletic shoes are often the footwear of choice when we plan to be on our feet for extended

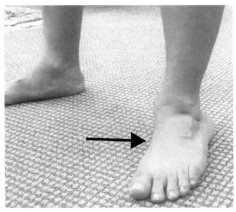

FIGURE 43A

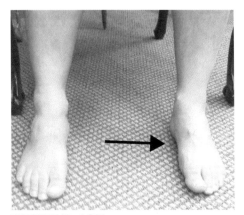

FIGURE 43B

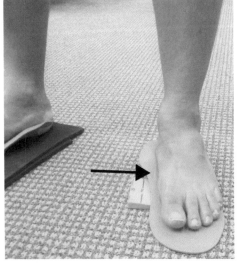

FIGURE 43C

As I've mentioned, most people don't have a rigidly fixed flat foot deformity, but rather one that changes from sitting to standing. Figure 43A shows a patient with a severely pronated foot when standing, yet this same foot can be put into a completely corrected position when sitting (Figure 43B). Figure 43C shows ideal alignment being created while being fitted for fully custom orthotics. Notice how this alignment is fully maintained even with complete weight-bearing.

periods. Many dress shoes now come with soft soles and removable inserts. Independent manufacturers have developed all sorts of additional soft supports for shoes that don't already have them, or to improve those that do. If you haven't already done so, simply look online and you'll be amazed at the multitude of vendors offering some sort of supportive foot device.

Nevertheless, a recent French study [4] showed that "77% of surveyed doctors and rheumatologists rarely or never use orthotics for osteoarthritis of the knee, and 88% rarely or never use them for osteoarthritis of the hip." Unfortunately, many physicians treating arthritis in this country don't use orthotics either… but they should…because foot orthotics most certainly have a much greater value than just comfort.

An article in the *British Journal of Sports Medicine*, [5] reported that, "biomechanical factors are increasingly being recognized as potential contributors to the cause and pathogenesis of knee osteoarthritis," and that "the correction of biomechanical variables in people with established knee osteoarthritis may delay the progression of the disease." *Orthotics work because they do just that…correct the mechanical factors that cause arthritis of the weight-bearing joints.*

As I mentioned earlier, in April 2009, fifty organizations gathered for a summit meeting on osteoarthritis. Their report made reference to the body of data regarding the positive, low risk effects of orthotics on mechanical loading of the joints [6].

A 2009 pamphlet from the Arthritis Foundation further validates the use of orthotics. Although the pamphlet still states that, "the cause of osteoarthritis is not known," and makes no mention of joint alignment in that regard, it does state that "assistive devices can help to decrease your pain and improve your ability to move," and that "shoe orthotics are examples of external supports that can help stabilize joints, provide support to joints, correct joint alignment or prevent further deformity of a joint."

Because of their effectiveness, orthotics are as commonly used by podiatrists as cortisone cream by dermatologists. Why? Because, as I have reiterated in this chapter, *they can create the optimal alignment necessary to decreases abnormal friction and pressure, which can ultimately destroy joints.* As we discussed in Chapter 6, custom foot orthotics not only control abnormal pronation and other foot abnormalities, but can also deliver needed changes for leg length discrepancies, as long as those changes are not too great.

Gross and Hillstrom, writing in *Rheumatic Disease Clinics of North America,* stated that, "Nearly all patients who have symptomatic knee osteoarthritis report that their pain is provoked by some kind of weight-bearing activity" [7, p. 756]. It is the combined forces of malalignment and gravity which creates the abnormal load on the weight-bearing joints that causes symptoms, and the joints' subsequent destruction. That's why canes are so popular... they relieve some of the load on arthritic joints. Orthotics work in a similar fashion, but far more effectively, by immediately relieving undue pressure at our foundation...like a new set of shocks and springs on an old car.

THERAPEUTIC EFFECTS OF CONTROLLING MALALIGNMENT

Reviewing what I mentioned in Chapter 2: Much of what I and others have learned about controlling symptomatic, abnormal alignment of the weight-bearing joints comes from treating runners. The custom foot orthotics I was given back in 1972, did more than just eliminate my symptoms. They made me realize how important proper foot position is in controlling pain in the weight bearing joints that the foot supports.

As you saw, when I began treating runners and other athletes by controlling abnormal pronation, I was amazed to see that in many cases, simply optimizing their foot position was all that was needed to eliminate painful symptoms in their knees and hips. This was true of conditions like runner's knee or *chondromalacia* (pain in the front of the knee often associated with sports), iliotibial band syndrome (pain on the outside of the knee or hip), and others. The principles of alignment that I learned and then applied in treating their ailments, as well as in increasing their sports performance, is the basis for the success I now experience in treating arthritic patients.

I think it's important for you to see some of the impressive data that proves the efficacy of foot orthotics in controlling abnormal function, and thus reducing arthritic symptoms. So, just as I did in Chapter 3, I'll share some of this information with you. What's listed below is just a mere sampling of the available, supportive documentation.

Since much of my belief and subsequent use of orthotics began with runners, I'll show you some of that data first. Comments will be made to help clarify this important information. *Remember, these very same mechanical principles can be applied in helping you with your arthritis.*

Williams et al. [8] reported significant differences and thus improvements in tibial torsion and knee abduction, i.e., leg and knee position, with orthotics. Excessive pronation causes an inward twist or turning of the tibia (inside lower leg bone) and knee that is often responsible for arthritis in this joint. What these researchers found, was that by controlling excessive pronation in runners with orthotics, there was a decrease in excessive destructive motion in the legs and knees of those tested.

MacLean et al., [9] demonstrated improved knee function, even in runners without any symptoms at all. Many people feel that if they don't have any symptoms, they must not have any underlying problems. Yet we know this is far from the truth. Many patients who have heart attacks have never had any symptoms either. This study shows that even those with no symptoms can be made to perform more effectively, more optimally, and thus decrease joint stress.

A 2003 study on runners [10] at the Human Performance Laboratory, at the University of Calgary, showed that orthotics decreased excessive pronation, increased vertical loading, and external knee rotation. In essence, this study showed that vertical alignment was enhanced, thus decreasing joint stress. Similarly, improved function has been demonstrated by using foot orthotics in those with knee osteoarthritis [11, 12].

Gross and Foxworth's [13] literature review and personal experiences, showed a reduction in patellofemoral (kneecap) pain when abnormal pronation and other malalignment issues were corrected with orthotics. A far more extensive literature review conducted by Marks and Penton, [14] found strong evidence for a reduction in symptoms and improvement in biomechanics in those with knee osteoarthritis, when orthotics were utilized.

When foot orthotics are added to other treatment modalities, results are enhanced. Krohn [15] demonstrated improvement in those with knee osteoarthritis by combining foot orthotics with knee bracing. In a study conducted by of Saxena and Haddad [16] 76.5% of their patients with patellofemoral (knee) pain syndrome had a reduction in symptoms following the use of orthotics.

A study by Rubin and Menz [17] was most impressive. They showed that although a greater reduction was obviously achieved in less severe cases, at 6 weeks, *all subjects with knee osteoarthritis had demonstrated some reduction in symptoms when fitted with orthotics.*

Johnston and Gross's [18] study revealed significant improvements in patellofemoral pain syndrome cases, in pain and stiffness, in just 2 weeks after being treated with custom foot inserts.

Keating et al. [19] found, as I have...*that sometimes even in severe cases of knee osteoarthritis...where there was a complete loss of joint space and bony erosion, some patients showed improvement in symptoms with orthotics.*

But these improvements are certainly not limited to just OA of the knees. Dananberg and Guiliano's [20] study, showed that *patients treated with custom foot inserts experienced more than twice the alleviation of chronic low back pain, for twice as long, when compared to traditional modes of therapy.*

Children with juvenile rheumatoid arthritis have been shown to have significant improvements in pain, level of disability, and improved function, with custom foot inserts [21].

A number of studies have attested to the efficacy of custom foot inserts in treatment of adult rheumatoid arthritis [22, 23, 24, 25, 26, 27, 28, and 29]. As mentioned, this is a far more severe and destructive form of arthritis than osteoarthritis. Although not primarily *due to* malalignment, as is osteoarthritis, this disease alters structural mechanics, making symptoms much worse. Orthotics are effective because they help control the abnormal function that this and other diseases cause.

The effectiveness of custom foot inserts on other conditions such as hemophilia (a blood disorder that can cause joint damage), has also been shown [30]. Beyond improvements in mechanics and symptoms, custom foot orthotics can decrease the need for NSAIDS (nonsteroidal anti-inflammatory drugs) and other medications [4, 31]. A study by Gélis et al. [4] for example, found that those treated with foot inserts for knee osteoarthritis consumed fewer NSAIDS than the placebo group for up to 2 years. *This is quite important, as it shows that foot orthotics...a simple, noninvasive form of therapy....can greatly reduce the amount of drugs and their subsequent, sometimes dangerous side effects.*

Brouwer et al., [31] also found that merely strapping the subtalar joint to prevent excessive pronation, had significant improvements on osteoarthritic knee pain, during ambulation as well as when at rest. As I've mentioned, the subtalar joint is the structural core of the body and excessive pronation at this joint is the primary cause of arthritis of the weight-bearing joints. *This particular study showed that even when this important joint is just taped... significant reductions in knee pain can be achieved.*

ADDITIONAL BENEFITS

Orthotics, as mentioned earlier, can also benefit other conditions such as diabetes. By altering function, orthotics can decrease areas of friction and pressure that might otherwise result in the tissue breakdown and subsequent ulcerations so feared by people with this disease. Orthotics are also often used in the treatment of neurological cases such as strokes or multiple scleroses, to add stability and balance, and cushion fragile skin during walking.

For example, a patient that has had a stroke on their right side must use their left foot much more to push off from when walking. The ability to do this is dependent to a large extent on their left foot being a stable, rigid lever. A good foot orthotic can provide this and can make quite a difference in their ability to walk.

Patients with obesity, hypertension, high cholesterol, heart disease and other medical problems that plague our society, are instructed to walk or do other types of exercise...quite difficult, if not impossible to do...when knee or other joint pain is exacerbated by such activities. By enabling more pain free joint motion, orthotics can help many of these patients stay more active. They therefore, have a very important indirect effect on these diseases.

Optimal alignment can significantly increase sports performance too, even as you age. Endurance, speed, balance, agility, and strength can all be positively affected by using orthotics. I'll talk more about this important benefit in chapter 8.

MAKING THE BEST FULLY CUSTOM ORTHOTIC

To truly alter function and be maximally effective, a foot orthotic must be accurately made to conform to the specifications of a precise prescription. After all, missing a putt by an inch is still a miss. That prescription is based on a thorough medical evaluation which has two basic parts: the structural examination and the corrective orthotic assessment. Then, a very accurate mold must be taken of the foot in its ideal functioning position. I also generally schedule a number of follow-up examinations to determine the level of comfort and degree of correction obtained.

THE STRUCTURAL EXAMINATION

There are many aspects to a structural examination, some of which include a gait analysis (watching the patient walk), assessment of leg-length differences, range of joint motion, flexibility, strength, and changes in the patient's structure when sitting and standing. This latter part of the examination is extremely important and helps determine the influence that structural abnormalities have on the weight-bearing joints.

Note that some aspects of my structural examination and orthotic assessment have been discussed in Chapters 5 and 6. Highlights are reiterated here for clarity.

CORRECTIVE ORTHOTIC ASSESSMENT

The structural examination is followed by an assessment of the orthotic correction required. Exacting measurements are the starting point. If you have ever had prescription eye glasses made, being fitted for fully custom, optimally aligning foot inserts should be very similar.

I ask the patient to stand on a sample set of foot orthotics. These have a specific baseline degree of correction which my lab has recorded, and which is used as a starting point in making final determinations. Then a set of wedges, with specific degrees of marked corrections, is used for both forefoot or front, and rearfoot or back measurements on each foot. These are inserted medially or laterally under the sample devices to increase or decrease corrections from the baseline.

Various leg length corrections are also assessed. This is done by having the patient stand on sheets of hardened rubber or other substances which are placed under the sample insert on the short side. I begin by using quarter inch thick corrections. Observations are made to determine if optimal positioning of the foot (subtalar neutral) is obtained (See Figures 44A, and B).

The patient is also asked to squat and while doing so, careful determinations are made to see if their knees are tracking more vertically, and whether painful symptoms are reduced with the contemplated corrections (See Figures 31A, B).

The patient's age, height, weight, type of shoe, specific sport or activity, level of expertise in that sport, severity of problem and other factors, will help determine which type of device to recommend, and the specific materials to be used. The prescription that is sent to the lab will include all of these specific measurements and findings.

TAKING A FOOT IMPRESSION

Although opinions differ as to the ideal way to take a foot impression, *the author believes it is critically important to do so with the patient non weight-bearing.* Only in this way, can the practitioner capture the foot in its optimal, subtalar joint neutral position. The entire shape and structure of the foot changes when the forces of weight and gravity are applied. Because of its profound importance, I want to be absolutely clear…in my experience, *ultimate correction cannot be achieved from a foot mold taken while the patient is standing in a weight-bearing position.*

To take a non weight-bearing impression, the patient must be lying down or sitting as opposed to standing or walking. While sophisticated pressure

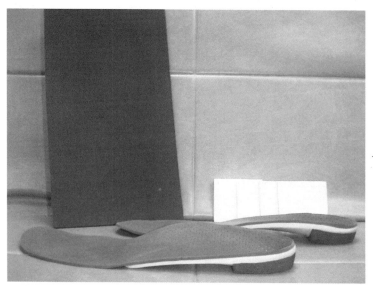

FIGURE 44A

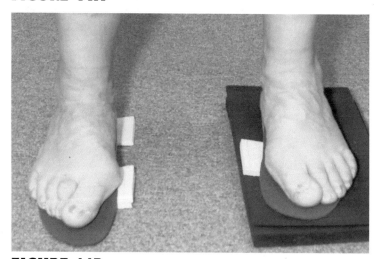

FIGURE 44B

Figure 44A shows some of the simple tools I use for assessing the specific corrections which will be included in the orthotic prescription: A sample set of orthotics with a predetermined amount of correction, ¼" hardened rubber sheets used to evaluate corrections for leg length discrepancies, and a series of incrementally marked wedges, to increase or decrease corrections to the sample orthotics. Figure 44B shows a patient being evaluated with these devices.

plates and various gait analysis recorders may look impressive, and some certainly have importance in evaluating patients, others are frankly gimmicks, resulting in ineffective end products.

In my experience, *properly capturing the foot in its ideal functioning position with a plaster slipper cast, simple foam box impression, or electronic scanner… will net far better results* (See Figures 45A,B,C,D,E, and F).

The art of casting the foot in its most effective position, called subtalar neutral, was popularized by Dr. Merton Root, in California, in the late 1960s and early 70s. Despite my extensive training, I knew that if I couldn't capture the foot in this critically important optimal position, my results would be significantly comprised. I was therefore fortunate indeed, to have been able to seek out and study with the esteemed Dr. Root in 1972.

Making slipper casts out of plaster of Paris was an art form to me. Although they were difficult, messy, and in the beginning we even had to mix our own plaster, I used them for many years. Today I can get even better results using a simple box of compressible foam, much like that used by florists to stabilize flower arrangements.

Knowing exactly how to hold and position the foot, and where to apply pressure to capture this ideal position, are of the utmost importance when taking a mold of the foot. When having orthotics made, one cannot begin to expect similar results just because two practitioners *supposedly use* the same technique, anymore than you could assume that surgical results will be exactly the same because two surgeons use the same procedure. The specific technique they were taught, and the level of expertise with which that technique is routinely implemented makes quite a difference. This is one of the many reasons patients are often dissatisfied with their orthotics, and incorrectly assume that "they"…meaning *any* insert…doesn't work for them. Remember too, that there are many other considerations that can be included or excluded from the final orthotic prescription and that these too, vary from practitioner to practitioner.

In summary, the accuracy of the technique used to capture the foot in its ideal position is the basis for, and directly related to, the degree of correction the

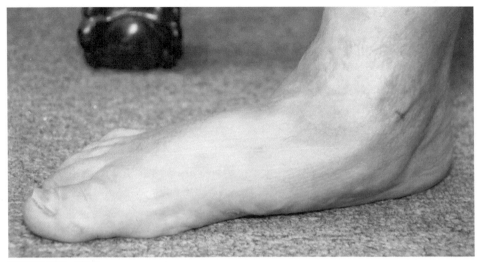

FIGURE 45A

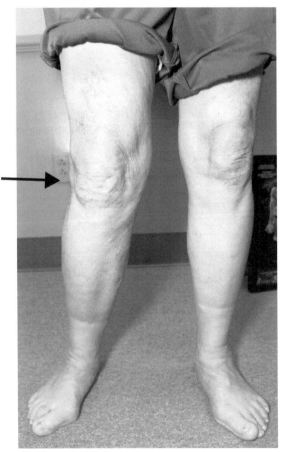

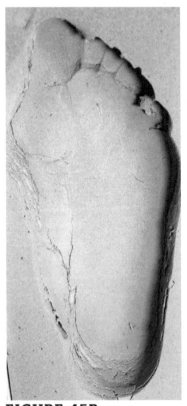

FIGURE 45B

FIGURE 45C

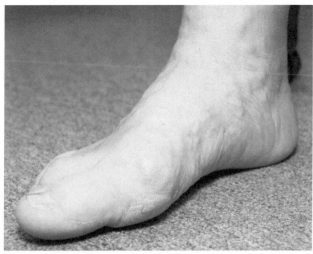

FIGURE 45D

FIGURE 45E

Figures 45A and C show a patient standing with a maximally flattened or pronated foot. An impression taken with the foot in this position, as seen in Figure 45B, often results in a device that actually holds the foot abnormally pronated, and therefore the knee, in an abnormal, internally rotated position. Figure 45D, shows the same patient's foot while he is sitting in a non weight-bearing, far more corrected position. When taking a foot impression in this position, seen in Figure 45E, an optimally corrected orthotic can be made which can correct and align the position of the foot, knee and other weight bearing joints, as seen in Figure 45F.

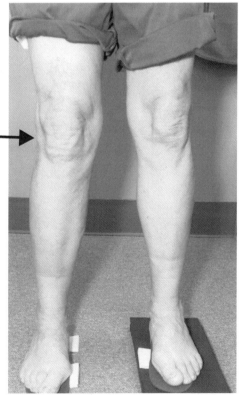

FIGURE 45F

orthotic can provide, and thus the level of improved function and decreased symptoms one can ultimately expect.

Once the cast has been taken, the patient is treated to alleviate symptoms. This may include teaching appropriate stretching exercises, providing temporary padding to decrease excessive pronation or partially correct a leg length discrepancy, administering injections, and other types of therapy. In the meantime, the cast or foot impressions are sent to the lab where the orthotics will be made. I am currently working on finalizing the use of computerized scanning in my office, so that devices can actually begin to be fabricated while the patient is being examined.

When the finished orthotics arrives, the patient is reevaluated to see if optimal correction has been achieved. Wearing instructions are explained and a follow-up appointment is made. At that time additional corrections or adjustments may be needed.

This aspect of treatment is just as important as properly casting the patient. For this reason, the procedure of adding or decreasing specifically measured wedges under the front and back of each foot may be repeated a few times. The same is true for corrections for leg length discrepancies. But ultimate alignment must be secondary to patient comfort. Often, because patients have adapted to their abnormalities, or normal bone and joint contours may have changed due to arthritis or injury, gradual corrections are necessary. Some may not ever be able to tolerate exacting final corrections for the same reason.

If distance or other factors are an issue, patients can e-mail one or two pictures of themselves standing barefoot on their inserts. From these I can usually make the appropriate corrections, and then send their devices back to them. This method is often preferred by many of my professional athletes and busy executives that I treat from out of town, as it enables them to only have to visit me once.

PITFALLS IN TREATMENT

So if orthotics can be so effective…why don't rheumatologists and other doctors use them more regularly and pay more attention to the actual degree of correction made? Why aren't they part of the standard of care for patients with arthritis of the weight-bearing joints? And why are so many others dissatisfied with the orthotics they did get? You'll find some of the answers to these questions discussed below.

1. MANAGED CARE

Now that I've explained the detailed and precise methods I use, which include a thorough evaluation generally lasting more than an hour, and the necessary follow up examinations, you can see why it is virtually impossible for most doctors to do this in our present, time oriented, managed care environment. That is why the marketplace is flooded with *similar looking, but far less corrective and cheaper devices.*

Remember, while over the counter magnifying glasses may work for some, you cannot possibly compare these to precisely made, optimally corrective eyeglasses that are individually prescribed for each eye, by a physician who thoroughly examined you. If you have prescription eye glasses, you certainly understand the importance of making orthotics that are made to the very same type of exacting specifications.

2. MISSING THE POINT

Since the importance of properly aligned feet in arthritis of the weight-bearing joints is not yet fully appreciated or accepted by rheumatologists and others, this concept is often not taught or practiced. Until this changes, the impetus to provide optimally corrective orthotics as a routine treatment for arthritis will not be commonplace.

3. ABSENCE OF STANDARDIZATION

Although there can certainly be exceptions, if you're being treated by a board certified physician, you can generally assume that you are seeing a more qualified doctor than one who isn't. But the world of orthotics is unregulated. So products are often dispensed by unqualified individuals making unsubstantiated claims. Unfortunately, you really don't know what you're getting, regardless of how much you have spent, or who you have seen.

---

[9] As this is not a text for physicians, and because the actual technique would be better demonstrated, it is not discussed here.

This is especially true because of the prevalence of the many look-alikes, confusing terminology, and deceitful advertising that I mentioned. I am currently working on the establishment of standardized credentials for orthotic certification to help eliminate this situation.

4. LACK OF PRECISION

When engine builders are measuring tolerances to thousandths of an inch for better performance and greater duration, and races are won by hundredths of a second, shouldn't the precision of a foot orthotic be as exacting as possible? Unfortunately, many who dispense orthotics have a nonchalant attitude...nowhere near the concern for ideal correction, for example, routinely exhibited by those dispensing prescription eye glasses or prescription medications. Physicians must not only cast patients' feet properly but must also be willing to continually make the necessary corrections needed to achieve optimal alignment, much as the specific dosage of a medication is regulated and changed to affect its optimum result.

I'm a perfectionist and view this aspect of medical care the same way I have always regarded surgery...with pride in precision and accuracy. It is frankly

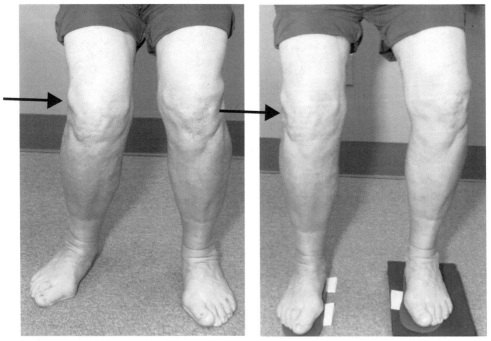

FIGURE 46B   **FIGURE 46C**

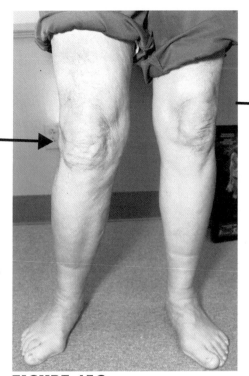

FIGURE 45C

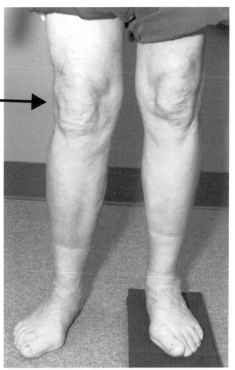

FIGURE 46A

Figures 45C and 46A (above) show the immediate difference in position of this patient's right knee when a lift is placed under his shorter left side. Notice in Figure 45F (right), the additional level of improvement in vertical alignment when sample orthotics and wedges are utilized. A simple way to assess how well a patient will ultimately function is to have them do a simple squat. Notice again in Figures 46B and C (left), the improvement in vertical alignment as soon as corrections are made.

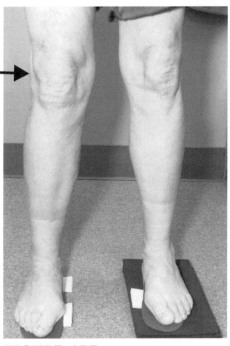

FIGURE 45F

a lot easier to give someone a soft device that's extremely tolerable immediately, than to try to optimally reposition poorly aligned joints while still keeping them comfortable. Like a fine piece of machinery, precision is the name of the game here.

I have reiterated the importance of precision a number of times in this chapter because it is the absence of this precision that prevents many arthritic sufferers from obtaining the pain relief they otherwise might so easily experience (See Figures 45C, 46A, 45F, 46B, and 46C).

5. INADEQUATE PATIENT EVALUATION

As you have seen, many important diagnostic findings must be incorporated into a good foot orthotic. Some patients who pronate excessively cannot tolerate very rigid devices, whereas heavy individuals need more rigidity to support their extra weight. Thick devices won't fit into running cleats or high heel shoes, and thin ones generally won't be supportive enough for long walks. People with a thinning of their skin and protective fat pads under the balls of their feet, and those with high arched feet, will need far more cushioning than those with more pronated or flattened feet. Patients with heel spurs may need special accommodations built directly into their devices to disperse pressure. Because so many other factors besides optimal alignment must be considered in making a finely engineered set of fully custom foot inserts, failure to perform a thorough examination and evaluation can result in less than optimal results. In fact…it may even result in complications.

6. IMPROPER CASTING

This is one of the primary reasons for ineffective, uncomfortable devices. How important is the impression a dentist takes for an inlay? How uncomfortable would a pair of eyeglasses be if you tried to wear them with the wrong prescription? If it's not precise, the device will never be comfortable regardless of how long you try to get used to it.

7. POOR PATIENT COMPLIANCE

Patients must be compliant when a physician is willing to do the serial corrections needed to maximize results. In most cases it has taken many years to create the pathology…it may take a while to correct serious problems and achieve full comfortable.

My patients realize that their fee is a "case fee," and includes all the follow up evaluations and adjustments to their devices until the desired effect is achieved. They also know that every time I send their devices back to the lab for correction, I incur additional costs…so if I'm doing so, it's for a very good reason.

Time is often a critical issue for both patients and physicians. As I mentioned earlier in this chapter, in many instances a simple picture of a patient standing barefoot on their inserts enables me to make the appropriate corrections, without the patient having to repeatedly come to my office. I also offer a "one day rush" service, so that when needed, patients can get their orthotics quickly.

8. HIGH COST

Optimally corrective, precisely made, fully custom foot inserts are expensive…and I think by now you can see why. As I have said, they should not be compared to most devices that seem or look the same and those that are typically covered by managed care contracts. Remember too, that what you are really purchasing is a precision prescription that creates the correction, the expertise involved in deciding which type of device is best for you, exactly how it should be made and many other factors, and not merely something that makes your shoes more comfortable.

When you compare the costs incurred and the potential risks of medications and surgery, with the powerful, long term effects of fully custom foot inserts, I believe there is no better value. In addition, a well made set of orthotics can last many years. Some of my patients have worn the same devices for more than 30 years, needing only a replacement top cover or other minor refurbishing.

In summary…
Remember that your feet are the foundation of your entire skeletal system. The position of your feet ultimately determines how you stand and function. Anything you place under your arch will usually help to some degree,

SELF-ASSESSMENT

How Can You Tell Whether Your Orthotics Are Maximally Corrective?

The key is to identify your optimal foot position ...where the bones of the talus and calcaneus are vertically aligned and stable. I have shown you how to see this both when sitting down and standing in front of a mirror (See the self assessments in Chapter 6).

If you're like most everyone else, and have feet that flatten somewhat when you are standing, do so, and attempt to perform a partial squat (See Figure 46B). You will feel the pressure on the insides and bottoms of the arches of your feet as they flatten, as well as on the insides of your knees as they turn inward.

While standing in front of a mirror, repeat this on your inserts. If they have any degree of rigidity and correction, you should feel them holding you in a more optimal position, and notice that you are more vertically aligned. Now place some small wedges on the bottom inner sides of your inserts just behind the big toe, and also where your heel meets your arch (See Figure 46C). Any wedge shaped material will work well.

If it now looks like your knees are even more vertically positioned over your feet and facing more outward, and that the bones of your feet seem more aligned, or that you feel more vertical, stable, balanced, are squatting more easily, and feeling stronger when you do so, then your inserts probably need additional corrections.

but precisely prescribed orthotics can have far more significant benefits. By properly repositioning and aligning the joints in the feet, these and all the weight-bearing joints they support (ankles, knees, hips, back, and neck), as well as their associated muscles, tendons, and other structures, will function in a more optimal position. This means reduced joint stress and therefore, subsequent joint damage. It also means increased stability and balance, decreased fatigue, and in most cases, reduced symptoms in your feet, ankles, knees, hip, back and neck. As you will see in the next chapter, it can also mean increased sports performance...sometimes at any age!

"*I had an injury to my big toe joint with subsequent arthritis, from years of playing semi-pro soccer. In 1982, Dr. Pack removed the bone chips and did a joint replacement. The joint implants in those days were made of plastic and I was told it might only last a couple of years. But he designed some special foot inserts to control my alignment and prevent stress on that new joint. Because of this, it never needed replacing and has served me well to this very day. I am still wearing the same set of inserts too! The implant worked great from the onset and I could resume full activity with no pain whatsoever. After the surgery, even with all my old injuries and arthritis, at 52, I was able to run again! What a feeling! I even entered races and ran until I was 72, when old back injuries caused me to stop running. But now 80, I am still very active.*"

Walt Micheles

Former semi-pro soccer player and choral director

CHAPTER 8

Finding the Fountain of Youth

INCREASING PERFORMANCE at ANY AGE

- Age doesn't always have to mean decreased performance.
- Since we're not born perfect, everyone who engages in sports, regardless of how gifted, has structural abnormalities.
- Sports performance is most effectively enhanced if optimal alignment is created before specific techniques are taught. This is a missing link in training athletes.
- The two primary aspects of alignment that are critically important in increasing sports performance are limb length equalization and foot position (stabilization)
- To a large extent, foot position determines how you start, as well as how fast and efficient your journey will be...how and where you will end.
- Changing foot position is one of the most powerful, dramatic, and effective ways to immediately enhance performance at any age.
- Regardless of brand or quality, shoes alone cannot offer optimal performance without a custom foot orthotic.

"I am in a very unique position to judge Dr. Pack's work. Personally, as a former competitive weight lifter and Swiss Army Officer, I had caused a lot of damage to my joints, resulting in arthritis, with lower back, calf and foot pain, that would not allow me to be active any longer.

Thanks to Dr. Pack, I now run 5 miles every day without any pain whatsoever. I subsequently flew my mother over from Switzerland who has suffered for years with near crippling arthritis of her feet. She too is now totally pain free and can enjoy her life again.

As a sports video analyst, I have in great detail, recorded, analyzed and interviewed many of the greatest athletes in the world, some of them senior athletes, both before and after seeing Dr. Pack. No one has had the unique opportunity that I have had to study the effects of his work.

What I have been able to show on video is simply amazing. The Olympic Gold and Silver Medals his athletes have won are testimony alone to the work he has done for many of them; speed, balance, alignment, power - all markedly improved, even in older athletes thought way past their prime. His work is truly unique and nothing short of genius.

Victor Bergonzoli

President and CEO, Dartfish, the world's leading company in sports video analysis. Dartfish was used by 95% of North American Medal winners in the 2010 Vancouver Olympics.

You will now see that the very same structural problems (we have been discussing) that causes arthritis of our weight bearing joints, greatly affect sports performance...at any age.

It was 2003, and as a faculty member of the United States Sports Academy, I was attending graduation ceremonies. There I met Roald Bradstock, who was being honored as the Sports Artist of the Year. Bradstock had set the world record in the javelin in 1984. Now in his forties and wondering whether his better days in sports were long gone, he was spending more time on his second passion...sports art. But as we spoke I could see the fire in his eyes and knew that his true love was still athletic competition. So I asked if he'd like my help in recapturing the performance of his youth.

After examining him, I smiled and said in a most lighthearted way, "the reason you can't throw well now is not because you're old and bald...you're a structural mess!" He obviously wasn't amused, and found it hard to understand how he could have set the world record if this were indeed true. I explained that when we're young we can compensate for our structural abnormalities, but that this becomes more and more difficult as we age. Although I obviously couldn't make him younger, I told him that I thought I could give him an edge he hadn't had even in his prime.

With great disbelief, the Brit decided, "to give it a go." In 2004 Roald Brad-stock became the first man *in any age group* to qualify for the Olympic trials...the oldest man ever to do so in this event. He has since gone on to set many more records and now hopes to compete in the 2012 summer Olympics in London. He will be 50 years old and dreams of winning again before his home town crowd.

Jerry Caldwell, a retired IBM executive and an avid tennis player, was in his sixties and could no longer play when he came to see me. After treatment, he went on to win the U.S. Tennis Association's Senior National Championship in 1986. Now 75, he still actively competes, beating opponents a fraction of his age. Jerry has been inducted into both the Georgia Tennis Hall of Fame and the Southern Tennis Hall of Fame.

Kurt Strater, has been skiing all his life but never expected to continue enjoying his passion well into his senior years. Yet at 82, he can still easily ski black diamond slopes with his grandchildren...even hitting some moguls from time to time. The knee pain he had experienced and attributed to osteoarthritis is now totally gone and he says that he "can actually turn faster and stronger now."

SO WHAT'S THE SECRET?

We're all too well aware of the affects of aging. Thinning hair, dry wrinkled skin, softening of our bones, generalized aches and pains, hormonal changes, a tendency to gain weight, and loss of memory, muscle mass, reflexes, and balance are just a part of a long list of disagreeable, aging characteristics. So it's no surprise that we also simply attribute our loss of sports performance to getting older.

As you have seen throughout this book, despite current thinking, age is not the primary cause of osteoarthritis of our weight bearing joints. Well, the same is true for sports performance. You see something is missing in the way we train athletes...even the greatest athletes in the world. A factor that when implemented, does more than just decrease arthritic symptoms, but creates an edge...at any level... and sometimes at any age. An edge that can often compensate a great deal for what aging has appeared to take away.

TRAINING AND STRUCTURAL ANALYSIS

Training any athlete usually begins with a medical clearance. At lower levels, or for young children, a note from one's physician might be all that is necessary before starting a program. In the pros and for Olympic athletes, much more detailed examinations are customary. Trainers and coaches will then routinely evaluate these premier athletes by testing them before training begins. Stability, balance, strength, speed, coordination, and flexibility are all among the things that are assessed. Depending on the goals of the athlete and the expertise of their trainers and coaches, nutritionists, sports psychologists and other experts may be called upon. Video analysis and other means of sophisticated technology may also be utilized. Then, based on their findings, the trainers and coaches will try to correct any deficiencies the athlete may have. This will include strengthening weak muscles, and improving flexibility.

Much emphasis is also placed on improving technique...such as holding one's arm in a certain position when hitting, or landing on a particular side of the one's foot when running. If you wanted to learn to play tennis for example, you would probably seek the advice of a professional who would recommend a particular racket and some lessons. The pro would watch you hit and then begin the lengthy process of teaching you proper form and technique.

Certainly, better techniques improve performance, and trainers and coaches deserve credit for doing so much to accomplish this. But remember...since none of us are perfect...*every single person who engages in any sport, no matter how gifted, has structural abnormalities that will make learning more difficult, decrease performance, increase the risk of injuries, and can later cause arthritic problems.* And make no mistake...yes...even the greatest athletes in the world are far from perfect. Many have serious structural issues but have learned to compensate for them.

Left untreated, these performance-robbing structural defects prevent all of us from reaching and sustaining our highest levels of functioning and therefore, achievement. They are often the reason someone doesn't make a team or win their event. Structural problems decrease the amount of time we can maintain a high level of performance. This can mean a shortened, much less lucrative career to a professional player or the lack of an active lifestyle for the average person. And when age and or arthritic problems, and former injuries, are added to long existing structural abnormalities, sports performance is further compromised.

Yet these structural abnormalities are rarely if ever fully recognized or optimally addressed. Golf, a game played by nearly 30 million people in the United States, many of them elderly, is a prime example of this. And the scenario presented below, occurs on a daily basis in almost every facility around the world, regardless of how renowned the institution, or the level of expertise of the teaching professionals.

If you want to begin playing properly, and "do it right," you are generally told to get a good set of clubs, have them fitted, take lessons, and then practice a great deal. If you want to go a step further, you can have some really sophisticated testing done. After being hooked up to numerous electrical connections, and having a state of the art video analysis recording, the problems with your golf swing will be very accurately assessed...which I think is great. But then... *your clubs will be expertly fitted to the abnormalities that were found in an attempt to help you maximize your performance.*

Although generally accepted and seemingly logical...this frankly makes no sense to me. Think about it....if you had a car with a bent frame you could tell everyone to sit on one side. But wouldn't it make more sense to fix the frame? If you are a right handed golfer with a longer left leg, you will never, ever be able to fully follow through on your swing, no matter how good your clubs or instructor. Wouldn't that be something that would be beneficial for both you and your instructor to know *before* you began taking lessons, practicing or competing?

And think about this…while some might argue that longer courses are the reason, despite oversized drivers, technologically advanced golf balls and shoes, irons made of space-age materials, state of the art putters, better conditioned courses and players, highly technical video teaching methods…even better tees…the average golf score hasn't changed since the early 1900's! If technology…which is what we continually focus on…was the answer, then how come this is true? *Because the only thing we have not corrected is the structural abnormalities of the golfers themselves!*

Neglecting to find and correct the inherent structural abnormalities an athlete (and all of us) has *first…before training them…* is like a physician prescribing medication before examining you and knowing what your illness is.

Contact sports are another example of the importance of evaluating athletes structurally prior to training or playing. In games like football where an apparent hit certainly seems like the cause for an injury…and of course, would be the *precipitating factor…* underlying structural problems may be the *predisposing factor.*

A talented quarterback for example, takes a bad hit and sprains his right ankle. Obviously, his being tackled caused his injury. But the fact that he may have had a short leg on that side, and always walked and ran on the outside of his right foot in an attempt to raise that shortened side, may very well have initiated his injury.

Many an athlete that thinks they have "weak ankles," has this type of unidentified and usually untreated problem; a problem that if treated properly, (sometimes with nothing more than a heel lift on a structurally shortened side), may have prevented this injury, saved a career, and helped them avoid an arthritic ankle.

Because I have subspecialized in a number of different aspects of medicine and surgery, I not only treat young athletes with such injuries, but see the end results...elderly patients with severe subsequent osteoarthritic ankle changes, due to years of abnormal functioning with what they were once told was a "weak ankle." Often a history of recurrent sports injuries is recounted.

The following is a case in point. It emphasizes that a simple structural evaluation could have saved this patient many years of suffering and changed his life. Remember, sports performance entails far more than just performing well...but how long a high level of performance can be maintained without injury. It further illustrates that often, even in very severe cases such as this, when it might seem that nothing can be done, simple corrections that specifically address abnormalities in alignment and function, can have remarkable results.

CASE PRESENTATION

Mr. W was a great high school football player who experienced recurring right ankle sprains, like the example discussed above. Improperly diagnosed and treated repeatedly, he subsequently dislocated the same ankle. His football days over, he spent the next 55 years walking and using this horribly malaligned ankle joint. When he could not tolerate the pain any longer or continue his daily farm work, Mr. W pursued surgical correction. But because of the severity of his problem and poor overall health, he was no longer a surgical candidate...he had simply waited too long. As a last resort, he sought my help (See Figures 47A, B, C, D, E, F and G).

Although Mr. W's case might frankly appear hopeless initially, after examining him I felt three important aspects of his problem could, and needed to be addressed: first, his much shorter (structurally shorter) right leg, which I believe precipitated his initial injury; second, repositioning of his right foot under his ankle (nonsurgically); and third, limiting the small amount of remaining ankle motion he had (see Figures 48A, B, and C).

If you do not have a joint, you cannot have joint pain. Pain is greatest when bone rubs against bone, in the absence of cartilage, its normal protective covering. That's why sometimes arthritic joints that do not appear as severe as others, may hurt worse than those that have no motion at all. When motion is no longer present, pain usually abates. This obviously simple scenario is the reason arthritic joints are sometimes surgically fused. Since surgery was not an option and the remaining motion Mr. W had was quite painful, a custom made ankle brace was utilized to limit that motion. This combined with the shoe modifications that addressed his structural issues resulted in a very satisfactory result (See Figures 49A, and B). Mr. W was able to carry on his daily activities with almost no discomfort.

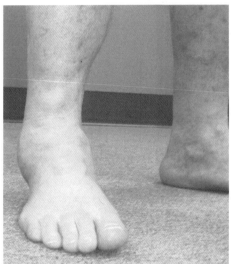

FIGURE 47A

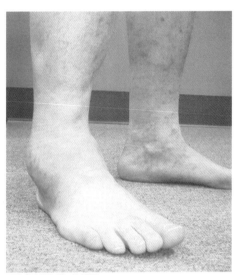

FIGURE 47B

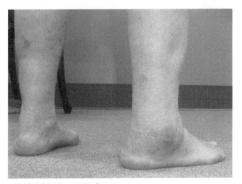

FIGURE 47C

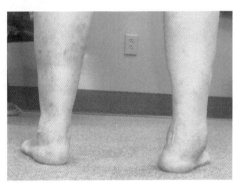

FIGURE 47D

Initially due to an apparent simple ankle sprain, improper treatment and subsequent injuries has resulted in the severely arthritic right ankle seen here...one that is poorly aligned, unstable and extremely painful.

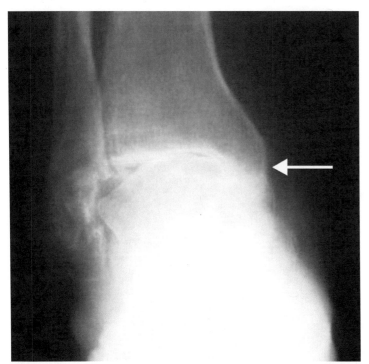

FIGURE 47E

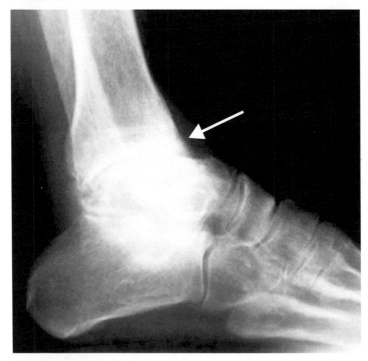

FIGURE 47F

ALIGNMENT: THE MISSING LINK IN SPORTS PERFORMANCE

Proper form is one of the primary goals of all trainers and coaches for any athlete. That's why technique is so emphasized in training. But initially evaluating athletes structurally for abnormalities like leg length discrepancies and improper foot position, and understanding the role these factors have on improper alignment and thus form...and subsequently on performance... is not commonplace.

Sports oriented Chiropractors and physical therapists appreciate alignment and wouldn't exist if there weren't so many players in need of their services and benefits didn't ensue. To be sure, athletes also recognize the importance of alignment. Golfers, sprinters, and other athletes take great pains to position their feet properly before they compete. *The missing factor however, is the creation and maintenance of optimal alignment, beginning with the athlete's feet, ideally before training, and on a more permanent basis than adjustments usually provide.* Appreciating a problem is one thing... optimally fixing it before hand to prevent injuries and increase performance is another entirely.

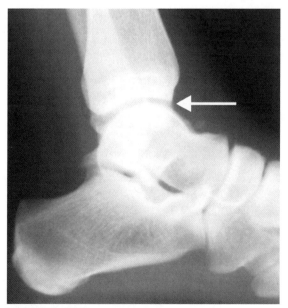

Radiographs of both the front (Figure 47E) and side view (Figure 47F) of his right ankle show obliteration of his ankle joint space, commonly referred to as "bone on bone" arthritis. Figure 47G is a lateral view of a normal ankle showing a fully open joint space.

FIGURE 47G

The two primary aspects of alignment that are critically important in increasing sports performance are limb length equalization and foot positioning. Although interdependent, it is ultimately proper foot position that is critically important. This consists for the most part of properly aligning the rearfoot; i.e., the talus over the calcaneus, and holding these bones in their best functioning position of subtalar neutral.

There is nothing I know of that can so often dramatically and immediately affect sports performance, as can this type of alignment...sometimes even in the elderly. That's because ideal foot positioning subsequently determines the alignment of all of the weight bearing joints the foot supports. This in turn decreases joint stress and increases almost every aspect of sports performance, including speed, power, strength, and balance. Certainly chiropractic adjustments and the work of physical therapists, kinesiologists and others, are also important in creating optimal musculoskeletal alignment. But it must all begin with a solid, stable base....our feet.

FIGURE 48A

FIGURE 48B

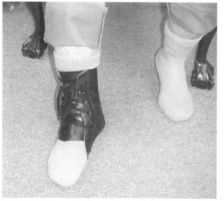

FIGURE 48C

Figure 48A shows this patient's work boots with a lift placed on the right side Notice that his right boot is sitting higher off the ground than the left. A lateral wedge was placed on the right boot as well, to help put his foot back under his leg in a much more aligned position, as seen in Figure 48B. Figure 48C shows a custom ankle brace that was also utilized to limit remaining ankle joint motion.

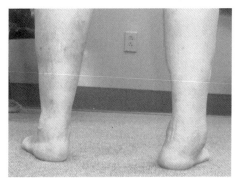

FIGURE 47D

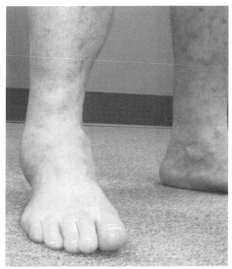

FIGURE 47A

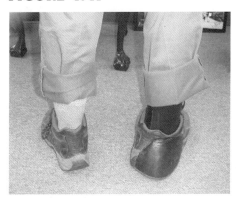

FIGURE 49B

FIGURE 49B

A dramatic difference in position of this patient's right foot can be seen with the corrections described above.

VERTICAL ALIGNMENT AND BALANCE

Balance, so important to golf and other sports, is defined as the "stability produced by the even distribution of weight on either side of a vertical axis" [1, p. 93]. This ideal of even weight distribution in humans begins and ends with our feet, which play a large role in determining the vertical axis of the spinal cord. As the great Sam Snead once said, "All good golf begins at the waist...all great golf begins at the feet." Unfortunately, as I've mentioned, very little is done to address this today.

Golf enthusiasts hear a great deal about Tiger Woods's exceptional degree of rotation; i.e., his ability to turn and fully bring his club back and then completely follow through. It is the envy of every golfer. I believe, this, to a large extent, is because his vertical axis is so straight. Whether he was born this way or has been corrected by someone like myself, I can't honestly say. What I can tell you, is that every time he picks up a club he has a huge structural advantage most of his competitors do not have.

The next time you have the opportunity to watch a golf tournament, rather than watching how Tiger looks teeing off, watch him stand and walk the fairways. You will see an individual that stands erect...straight and upright... and walks with a propulsive, powerful gait. This comes from more than just a good, confident mindset, but from great structural alignment. Almost no one else playing, stands or walks with such posture. By no means the only factor, his posture and the optimal function it provides has a lot to do with his winning edge.

To be clear...If I could be made laser beam straight, I still might never be able to play golf any better than I do now. That said, it is indeed a very important missing component in sports performance...again, at any age. And when all other factors are in place, by creating optimal alignment, I have fortunately been able to help a talented athlete make the pros or win an Olympic gold medal.

Generally, as indicated above, the rest of us golfers aren't as vertically erect as Tiger, often due to leg-length differences and other causes. Because of this, our plane of rotation is more angulated. So we are either swinging uphill

[10] Scoliosis and many other factors can also influence the position of the spinal cord.

or downhill all the time. In our efforts to overcome this, we develop what I refer to as, *score-inflating compensatory mechanisms*, which costs us dearly in terms of performance and injury.

Because of this, golfers often suffer from various musculoskeletal aches and pains which usually worsen as we age. Therefore, in an effort to simply make it to the next round, many top pros travel with an entourage, including chiropractors, physical therapists, and massage therapists. But why do these great athletes get so sore? They aren't playing football, or any contact sport for that matter. *Despite the current misconception that symptoms are simply inherent to their game, these athletes don't primarily get sore from swinging a golf club...but from doing so in poor alignment.*

More than the structural problems themselves, it is these compensatory mechanisms...like trying to forcibly swing through (over) a longer leg...that causes joints to move in a plane or position they aren't really designed for... that causes golfers to get sore. This is much like the example of the pitcher's elbow I mentioned earlier. To the contrary, proper alignment and balance not only increases performance, but decreases many of the aches, pains, and injuries caused by these repetitive compensatory motions.

"I am an avid golfer and have suffered for many years due to severe osteoarthritis of my hips, knees and back. I've had 2 left hip replacements and have 'bone on bone' arthritis of my left knee. I was told by two prominent orthopedic surgeons that this joint also needed replacement. On a 1-10 scale, my pain was an average of 8, and at times almost intolerable. This limited my activities significantly, sometimes making playing golf impossible. For an old athlete, this was quite disheartening. After seeing Dr. Pack my pain is down to a level 2 and continues to diminish. I cannot begin to tell you how grateful I am to be able to resume my activities, especially playing golf again and above all, now thanks to Dr. Pack.....avoid further joint replacement surgery.

Gary Coker, PhD

Formerly President, Brandon Hall College; Headmaster, Charlotte Christian School; and Chairman of Education, University of the South.

VERTICAL ALIGNMENT AND STRENGTH

Considered a major part of athletic training, much emphasis is placed today on the importance of core muscle strength. While certainly important, *strength involving the muscles of our weight-bearing joints is dependent upon, and maximized by vertical alignment.* So obtaining this alignment should be the *very first step* in optimizing strength without injury, and not the teaching of proper techniques to someone who is malaligned...which to some extent is all of us...or focusing on developing a strong core.

When replacing a tire, the first thing our auto manuals emphasize is the importance of being on a flat surface. Not only because the car may slip off the jack, but because the *power of that small jack is greatly maximized when it is vertically aligned.* And the same is true of us. I find it interesting that although these basic mechanical principles seem quite clear in many aspects of our daily lives, they are often left unconsidered in medicine and sports.

Anyone who has ever bench-pressed weights appreciates the importance of alignment. Your arms, which are generally non weight-bearing become so when lifting weights in this manner. Simply placing your hands unevenly on the bars makes it harder to lift heavy weights. Although often thought to be due to a weaker arm...and of course sometimes this is true...many an athlete would lift more weight with less chance of injury if their arms lengths were equalized. This same analogy holds true if squatting with weights on legs of uneven lengths.

VERTICAL ALIGNMENT AND SPORTS PERFORMANCE - EXAMPLES

A right-handed baseball player with a structurally longer left leg can't fully get through their swing, and will have to swing upward. Placing a correctly sized lift under the foot of a structurally shortened right leg, can have an often dramatic and immediate impact, on hitting in these athletes. The same is true for tennis players, golfers, and others.

As mentioned above, a golfer who is right handed and has a longer left leg can't turn easily to their left or follow through on their swing, whereas a golfer with a longer right leg can't bring their club back easily and will have a tendency to hit down on the ball. When you consider the fact that everyone has some leg-length discrepancy, is it any wonder golfers get frustrated? And I wonder just how many of you are saying at this very moment… "I'm glad I finally found out why my golf game is so lousy!"

A cyclist with a longer leg will often flatten their foot turning their knee inward on the longer side. This causes a loss of power and strength, especially needed for hill climbing, and can lower RPM's, resulting in a loss of speed. It can also increase the incidence of knee injuries on that side.

Leg-length differences make it harder for a kayaker to row straight, a swimmer to come off a turn evenly, or a horseback rider to stay balanced. Some riders' saddles actually become twisted from years of leaning to one side. Patients have also reported more injuries to their horses on the rider's shorter side because of the increased weight the leaning has produced.

THE ROLE OF THE FOOT IN SPORTS PERFORMANCE

Every sport begins and ends with the human foot…unless you're in a wheelchair…and even then, the wheels are your "feet," and *their* alignment and balance become critically important to your performance.

As I've said, many injuries can be attributed to poor foot position. So *optimizing how your feet function can be one of the most powerful and immediate means of improving your sports performance in any sport.* Yet despite the important role they play, *feet are perhaps the most neglected component of human sports performance, just as they are in all areas of medicine.*

The failure to understand exactly what the foot does, where and why it does it, and how it affects so much of what we do, has led to many of the problems athletes face. Ice skating is a prime example. If you've ever ice skated and rolled your feet inward, you, and many generations before you, were likely told that you had "weak ankles," and that your skates weren't tight enough. Like me, you may have been given a skate hook that enabled you to really cinch those skates so tightly around your ankles, your eyes felt like they were going to pop right out of your head.

As you may recall, the ankle joint is a hinge joint and can only move your foot up or down. So unless your ankle is broken or dislocated it can't possibly be responsible for your feet rolling in when you skate!

You now know that this excessive motion called pronation, actually occurs at the joint below the ankle, the subtalar joint, and would best be treated with a custom foot orthotic. I have designed orthotics for each and every sport... even gymnastics. Such devices, like the one for skating, specifically controls the abnormal motion at the particular joint at which the excessive movement actually occurs...the subtalar joint... and offers far more stability and performance than any tightening of the boot around the ankle could ever provide.

I therefore think, that the basic premise of our ice skates is incorrect and has led to the age-old art of making ice skates with boots above the ankle, with the intent of offering greater support for all those who roll their feet in. The result has been an increase in serious injuries to skaters' knees, especially when triple jumps and other complex movements are attempted.

To a large extent the same is true in skiing. Foot and ankle injuries have become infrequent in this sport, while the knee accounts for the great majority of problems. That's because when we wear a ski boot, we're virtually skiing in a foot and ankle cast which significantly limits motion in these joints, placing additional pressure on our knees. What happens to a machine if you limit the motion of an important moving part? That limited motion has to be taken up somewhere, causing additional stress on the other moving parts, and at some point the machine will break down.

As I've mentioned, I've been privileged to work with some of the greatest athletes in the world, including eight world record-holders in different sports as well as professionals in nearly every sport. Despite the often proven results that I have had in creating immediate enhanced performance in these athletes, simply with optimal foot positioning, this is still not common place. Olympic coaches have told me that they simply "did not have time to fool with the foot, because they had to focus on so many other aspects of training."

Research scientists were amazed when I suggested that to maximally affect pushing a bobsled, perhaps it was more important to focus on foot positioning than on how the athlete's hand was placed on the sled. After all, where does that push begin?

It's fascinating for me to watch the bobsled races in the winter Olympics and see these massive, powerful athletes wear what would appear to be ballet slippers, with no support at all, trying to push their sleds with severely flattened feet. With races being won by hundredths of a second, it's a shame these athletes don't realize that pronated feet like these, stay on the ground longer, costing them dearly in terms of time, and limiting the power of their push off.

Ask any athlete how much time has been spent properly aligning their feet and their structure and optimally holding that alignment, compared to teaching them correct technique, and you too, will be amazed. If you've been a competitive athlete, you already know this truth firsthand. I have been at this most of my life and have never, ever seen an athlete as a patient at any level, that was optimally aligned. And this includes some making tens of millions of dollars a year.

SPEED

Speed is one of the most important aspects of any sport. The fact is if you're slow, there aren't many sports you can excel in. But as in the example above, your feet can determine your speed to a large extent. And the length of your stride, the number of strides you take, and how long your foot is on the ground each time it strikes, are critical factors in that determination.

A blink is 1/1000 of a second. Decreasing the time the foot is on the ground by 1/5 of a blink, or 5/1000 of a second, for each stride in a 40-yard sprint, can reduce a 4.8 second runner's time to 4.5 or less. That can make the difference in an athlete making the NFL. The reason for this is because as I've mentioned, excessive pronation or flattening of the foot causes the foot to be on the ground longer, decreases the power of one's push-off, as well as increasing the stress on the rest of their weight-bearing joints (See Figures 50A, B, C, D and E)..

But it's not only straight-line speed. All aspects of performance are affected by a flattened foot. Excessive pronation delays the time it takes a baseball catcher to rise from a crouched position, giving runners on the opposing team quite an advantage. A flat foot is not as rigid a lever to push off from, making an athlete slower side to side. A baseball player with a good left but flatter right foot will be quicker to their right since the left foot is a better lever to push off from. Such a player would have an advantage in playing first base but a disadvantage in playing third…and yes… it can be and often is, *that* specific (see Figures 51A and B).

Tennis, basketball, soccer, and football players are affected the same way. And a player who is aware of this could look at the position of their opponent's feet and know which side they would have an advantage to pass on. For example, a basketball player with a flatter right foot and a good left foot who is playing against someone who also has a flattened right foot, would have a significant double advantage going to their right side, and to the left of their opponent.

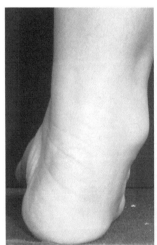 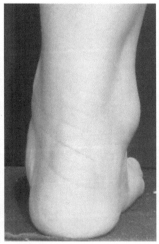 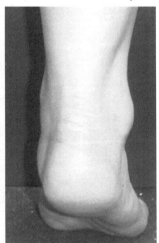

FIGURE 50A   **FIGURE 50B**   **FIGURE 50E**

The very same principles I have discussed with arthritis apply to sports performance. You are looking at the back of the right foot of a patient. The ideal foot hits the ground on the outside of the heel (Figure 50A) and then rolls inward, adapting to the ground until it reaches its optimal functioning position of subtalar neutral (Figure 50B). It then acts as a rigid lever, propelling you forward to your next step (Figure 50E).

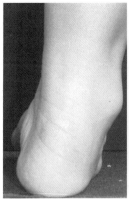

FIGURE 50A

FIGURE 50B

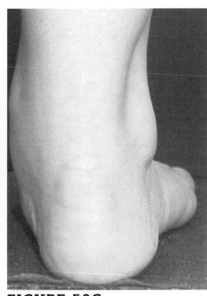

FIGURE 50C

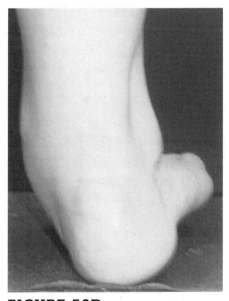

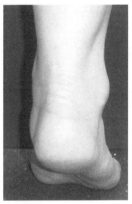

FIGURE 50E

FIGURE 50D

A foot that pronates excessively does not stop at this ideal position before pushing off, but continues to go through a triplane motion, rolling inward as the forefoot slides out laterally (Figure 50C) and turns upward (Figure 50D) before pushing off. These additional steps (larger pictures) cause the foot to be on the ground longer and results in a much less effective push off.

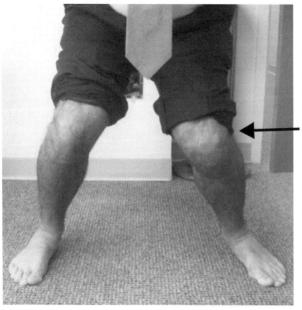

Once, a 4th round NFL draft choice and now golfer, this gifted, right handed athlete, was always faster to his left. That's because his longer left leg caused him to flatten his foot on that side in an attempt to shorten it. Because of this, his left foot was not as good a lever to push off from, compared to his well positioned right foot.

FIGURE 51A

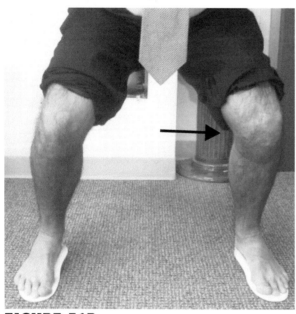

Now retired, he had significant arthritic pain on the inside of his left knee from years of functioning in this abnormal position. Once properly aligned and balanced with specially designed custom foot orthotics, which have a lift built into the right device to correct his shorter right leg, he no longer has knee pain. If still playing football, he would also be much faster to his right side. This right handed golfer now finds it a lot easier to play since he isn't constantly trying to swing up hill.

FIGURE 51B

Skiers who pronate will be slower because they are basically snowplowing all the time. This can be seen from the deeper groove made in the snow by the inside edges of their skis. Excessive pronation also makes it harder for them to push off and turn. So a skier with a flattened right foot will have a much easier time turning to that side, as he or she pushes off on the better positioned, more vertically aligned lever provided by their left foot.

And no....in most cases the ski boot inserts made in ski shops are not the same as those fully custom orthotics made by Podiatrists and others who specialize in sport orthotics. Remember that the device is only as good as the prescription, which is based, in turn, on a thorough musculoskeletal evaluation, and unless you have someone like a sports physician who is examining you and evaluating such things as leg length discrepancies, tight calf muscles etc., corrections for those components cannot be accurately included in the final device.

BALANCE

Balance is also greatly influenced by foot position. A foot that is optimally positioned in subtalar neutral offers more balance than one that is rolled to the outside or excessively flattens. This is especially true in sports that are at times performed on one foot...like ballet, pitching, pole vaulting, throwing the javelin, bowling, and gymnastics (especially on the balance beam). Well positioned feet are also important to golfers, who although continually redistribute their weight, depend upon both feet at all times for their balance. Optimal weight distribution is not possible, if for example, one leg is longer than the other...again, something we all have to some degree.

STRENGTH AND POWER

Like speed and balance, ultimate strength and power can also be greatly influenced by the position of one's feet. As you may recall, pronation is a triplanar motion. So when excessive, the foot doesn't just flatten and roll inward, but slips out laterally and rolls upward, basically sliding out from under one's body.

So people with flattened feet have a harder time pushing objects away because they have a weakened lever from which to propel. Because of this,

SELF-ASSESSMENT
Feel the Power

Note: Those of you with significant arthritis of your knees or other joints will obviously not be able to do this test.

Stand with your feet shoulders length apart. Roll them inward, flattening them. As you do this you will notice that your knees will face more inward, closer and towards each other. Now do a partial squat. Ask someone strong to stand behind you and push down on your shoulders as you try to push up into a straight position against their resistance. Repeat this standing on the outer edges of your feet while keeping your knees straight and further apart from each other. Don't let your knees turn inward or move closer towards each other. If you have a good set of foot orthotics, use them too. Both of you should feel a significant increase in power and strength with the other person having a much harder time resisting your efforts (See Figures 52A, B, C, and D).

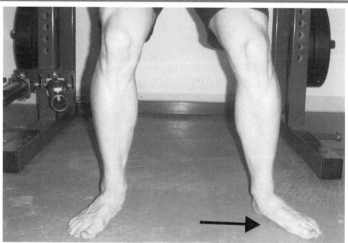

FIGURE 52A

This competitive weight lifter has significant pronation and a longer left leg...not a very good combination for lifting heavy weights without injury. Figures 52A and B show his feet collapsing causing his knees to come closer together when squatting with significant weight. Notice that his left foot is flatter and turns more outward, his

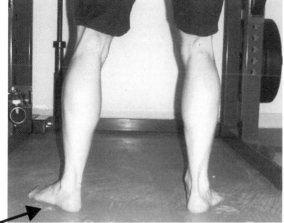

FIGURE 52B

left knee is more bent and turned in a greater outward direction than his right, in attempt to shorten his longer left leg. In Figures 52C and D, he is standing on temporary corrections, including a lift under his right foot, while being assessed for his custom foot orthotics. A dramatic difference in the positions of his feet and knees can readily be seen. Because his feet are now not sliding out from under his legs but pointing straight, his knees are well positioned over his feet, his left leg is no longer turning outward, his stance is wider and much more stable and all of his other weight-bearing joints are more vertically aligned. This corrected position will allow him to lift more weight with less injury immediately.

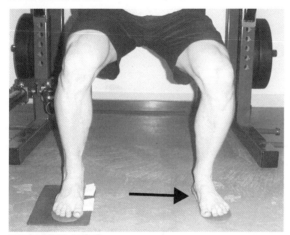

FIGURE 52C

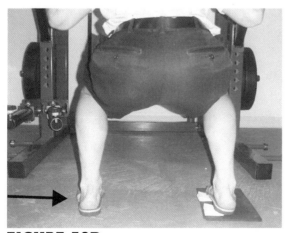

FIGURE 52D

trainers and top athletes often place their feet in a corrected stance position when asked to do a squat with weights. Football linemen, bobsledders and others can often really feel the difference optimal foot position has on power. Sprinters know the powerful advantage a starting block gives to their start.

Excessive pronation, which you saw causes instability and subsequent arthritis of the weight bearing joints, not only dramatically decrease the ability to lift weight, but can result in serious injuries in athletes.

This abnormal motion can also affect performance in less obvious ways. For example, when you bench-press weights you push off the balls of your feet. Like uneven arm length, strength can be lost if your feet are flattened. Although unrelated to pronation, the same is true if one arm is longer than the other. The power of a punch in boxing or in martial arts is also greatly affected by a flattened foot. Tennis strokes aren't as powerful, nor can a javelin, discus, or shot put be thrown as far.

Structural abnormalities combine to affect many aspects of an athlete's performance. For example, a right handed tennis player with a longer right leg and more flattened left foot, will have much more difficulty with their forehand than their backhand. The more pronated left foot will make them slower to move to their right side and the longer right leg will make it harder for them to turn to their right side as they try to hit the ball with their forehand. A right handed shortstop with a longer right leg and more flattened right foot will be hindered in both directions. Their longer right leg, making it harder for them to turn to their right and their more flattened right foot making them slower going to their left.

Reminder....structural abnormalities are certainly not the only reason sports performance may be compromised, no more than it is the only reason osteoarthritis develops. They are however, very important, usually neglected, often easily corrected contributory factors.

A WORD ABOUT ATHLETIC SHOES

Many trainers and athletes feel that a good pair of shoes is all that's needed on one's foot to make them perform well. While it's certainly true that advances in athletic shoe technology (especially those for running) have decreased injuries and led to better performance, runners and other athletes still experience the same problems they did when I first started treating them in the early 1970s. They're just running longer and faster before these problems occur.

Running-shoe manufacturing has become very sophisticated, offering more varieties than toothpaste. To help you decide which shoe is best suited to your particular foot and activity, many brands label their shoes by categories. These include those for pronators, or people whose feet roll inward and flatten, and those for supinators, who roll their feet toward the outside. But while these general classifications can be helpful, even the finest running shoes are generic. They're made by manufacturers to address "average" feet with "average" problems.

Even if you have a good idea of which category best fits your foot type, remember again, that you are not perfectly symmetrical. For example, one foot will always roll in more or one leg will be longer than the other. So with any pair of shoes, while one shoe may be ideal for one of your feet, it may actually cause a problem on your other foot. If your right foot for instance, supinates and rolls outward, you are lacking in pronation...some degree of which is necessary for proper functioning. A shoe that is made to control or limit pronation may very well cause your foot to be overcorrected and result in an ankle sprain or more serious injury.

Unless you're a podiatrist or some other type of foot expert, realize that you're more or less "shooting in the dark," when it comes to choosing appropriate athletic shoes. Remember too, that no *shoe can offer optimal performance and prevent injuries without a custom foot orthotic.* As you have seen, when made properly, such a device can correct many different types of abnormalities, and like a fine pair of prescription eye glasses, help you perform better with less strain.

Another important factor to consider is quality control. Today, like many other products, even top-brand athletic shoes are not made to the high standards we used to customarily expect and get. Manufactured overseas with major emphasis on cost reduction, athletic shoes may have many imperfections, including being made on a pronated last. This means that the shoe

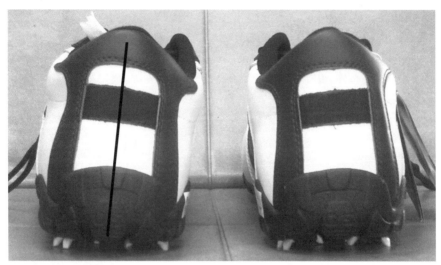

FIGURE 53A

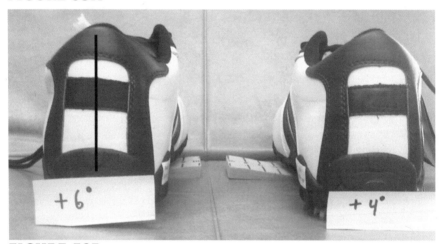

$+6°$

$+4°$

FIGURE 53B

Often, great athletes have their shoes custom made by the company endorsing them. Although hard to imagine, it is not uncommon for even these shoes to be defective and poorly aligned... something apparently unknown to both the professional, and their sponsoring company. Such shoes can actually decrease the performance advantage these athletes hoped to gain by wearing them. These pictures show a competitive golf pro's new shoes. As you can see, one of his shoes is off more than six degrees from vertical. If worn, they will actually hold this athlete in an abnormally pronated position. In essence, it would be like hitting with a different club than he thought he was using!

may actually be rolled inward before you ever wear them. Shoes like these can significantly decrease performance and increase the risk of injury.

When buying a new pair of shoes, always place them on a completely flat, horizontal table and make sure that an imaginary vertical line down the back of the shoe is perpendicular to the transverse surface (see Figure 53A). If they're not, find another pair. I have seen some top-brand golf shoes for example, that have been off more than six degrees right out of the box. This can make a significant difference when hitting a ball or performing in any sport (see Figures 53A and B).

OPTIMAL PERFORMANCE AND EVERYDAY LIFE

So, how does all this relate to aging?

As I've mentioned, when we're young, we sometimes learn to compensate for our structural abnormalities. In many instances great athletes aren't born more stable or better aligned that the rest of us, they simply compensate better. But as we age, we lose the ability to do this to some degree. This loss combined with the effects of years of malalignment, wreaks havoc on us; especially if our malalignment has caused injuries that have resulted in secondary osteoarthritis. And, while there isn't anything I or anyone else can do about certain aspects of aging (I'm still waiting for a cure for baldness), s
A 78-year-old, right-handed male golfer with a structurally longer left leg, who has been playing golf since childhood, probably has always had a hard time following through his swing. Although he may no longer have the flexibility he once had…or the grip, speed, or power…putting a lift in his shoe on his shorter right side, will most likely help him follow through much more easily, with a far greater degree of rotation…perhaps even more than when he was young. This should result in his ability to hit further and more accurately and at the same time, decrease the post game soreness he would otherwise likely have.

An older woman who has run most of her life and never had an injury, now finds she can no longer do so because of pain on the inside of her right knee. She is convinced that this is due to her age and years of running. But her left knee isn't any younger and unless she hopped on her right leg, that knee

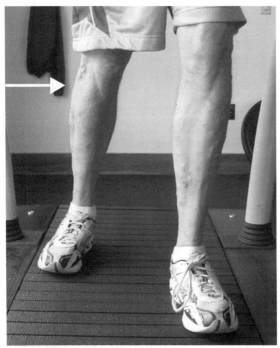

FIGURE 54A

As I showed in chapter 2, in each of these pictures the individual's right foot can be seen to pronate and their right knee turn inward. While these patients may be lowering their weight, cholesterol and blood pressure by exercising, doing so like this…in poor alignment…will cause excessive pressure on their right knee joint and cause it to deteriorate. This is true when using any piece of exercise equipment while poorly aligned.

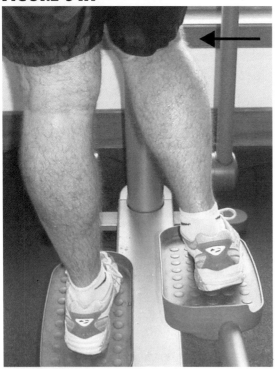

FIGURE 54B

received the same amount of stress as her right knee has all her life and is still pain free. If significant pronation or another structural abnormality is the primary reason for her right knee pain, correcting this problem may not only help her resume running, but may also help her perform better than she did when she was younger.

The elderly individual who has a hard time going for a walk with his dog or grandchildren, may not realize that his tight calf muscles, and not necessarily his age, may be largely responsible for the stiffness, fatigue, and the tripping he experiences. Once these muscles are properly stretched, he may have more motion than he's had in years.

I am blessed indeed. Nearly 64 years old, I continue to exercise daily, lifting weights, jumping rope and doing other strength and conditioning exercises, and sometimes training for my next marathon. Often this is done at an athletic facility with many of my senior citizen friends. I can't begin to tell you how disheartening it is to see these individuals who are working so hard to stay in shape, not realizing that they are actually making their joints much worse by exercising in poor alignment.

Most people use some sort of exercise apparatus for cardiovascular conditioning. If you're one of them, see if you can tell whether you are exercising in proper alignment or if your feet are pronating and your knees are turning inward. Perhaps ask a friend to look at you and see if they can tell. If so, be sure

"*I played hockey all my life and it has taken quite a toll on my joints. I feared my constant knee pain due to injuries and osteoarthritis would prevent me from being able to skate as I aged; after all, hockey really isn't a sport for seniors. But since you treated me, I can now skate totally pain free, and even have much better balance and feel than I used to have! If only I had met you 30 years ago. Guess I'll have to settle for being a senior star; there's no money or fame, but at least the beer is still cold. Thanks for helping me to continue to live my passion.*"

Tom Olney
2009 USA National Hockey Championship Team, seniors (over 50 years old)

to correct this before continuing to exercise. Remember: "Health is something you can go through on the way to fitness" (see Figures 54A and B).

In this chapter, I've emphasized that one of the best ways to improve sports performance is to improve structural alignment. Even if you don't consider yourself an athlete, this same principle can be used to substantially improve the quality of your everyday life.

Assessment and correction of structural abnormalities first, ideally before playing a sport, working out, or having a coach or trainer teach you specific techniques, combined with an appropriate physical regimen of strengthening and stretching, can have a profound effect on how you function at any age— often despite some arthritis that may be present.

In closing…

I hope this book has helped you understand the critically important role that structural abnormalities play in the prevention and treatment of osteoarthritis of the weight-bearing joints. Whether you are a world-class athlete or simply want to take a pain-free walk with your grandchildren…if I've encouraged you to seek treatment to correct problems like a flattened foot or longer leg, and in so doing you have alleviated some of your discomfort, I am truly happy for you and perhaps I will have accomplished my goal. And if this book helps change the way physicians and coaches look at those they are privileged to care for, I will certainly have fulfilled my mission.

Time is one of our most precious possessions. Thanks for giving me a little of yours and taking the journey!

Dr. Lou Pack

To learn more about the Arthritis Revolution, book Dr. Pack for a speaking engagement, or make an appointment, please visit www.drloupack.com

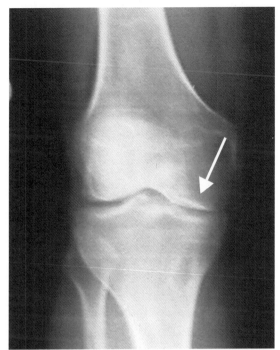

FIGURE 55A

RIGHT KNEE X-RAYS

62 year old female with chronic osteoarthritic pain in her right knee and left hip of 2 1/2 years duration. Her pain increased with activity, severely limiting her lifestyle. She was told that her only alternative was total joint replacement. The immediate and dramatic change in medial joint impingement seen here, occurred as soon as she wore the custom foot orthotics that we designed, which corrected her pronation and leg length discrepancy. No other therapy or medication of any kind was used. She now has been able to resume a far more active, less painful lifestyle.

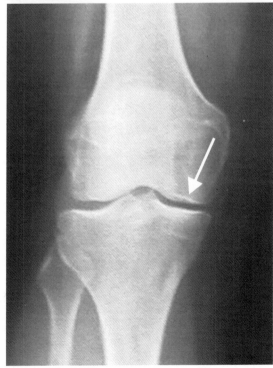

FIGURE 55B

REFERENCES

CHAPTER 1

1. Centers for Disease Control (CDC). (1994). Prevalence of disability and associated health conditions—United States, 1991–1992. Mortality and Morbidity Weekly Review, 43(40): 730–731, 737–739. Cited in Theodosakis et al. (1997), p. 2.

2. Yelin, E., & Callahan, L. F. (National Arthritis Data Work Group). (1995). The economic cost and social and psychological imnpact of musculoskeletal condition. Arthritis and Rheumatism, 38(10): 1351–1362. Cited in Theodosakis et al. (1997), p. 3.

3. National Institutes of Health (NIH). (2003). Consensus Development Conference Statement (December 8–10, 2003). NIH Consensus Develoment Conference on Total Knee Replacement. <http://consensus.nih.gov/2003/2 003TotalKneeReplacement117html.htm>. Accessed January 10, 2010.

4. Felson, D. T., Anderson, J. J., Naimark, A., et al. (1988). Obesity and knee osteoarthritis: The Framingham Study. Annals of Internal Medicine, 109(1): 18–24. Cited in Theodosakis et al. (1997), p. 236.

5. Sandmark, H., Hogstedt, C., Lewold, S., & Vingard, E. (1999). Osteoarthrosis of the knee in men and women in association with overweight, smoking, and hormone therapy. Annals of the Rheumatic Diseases, 58(3): 151–155. Cited in Theodosakis et al. (1997), pp. 236–237.

6. Arthritis Foundation (n.d.). What is osteoarthritis? Arthritis Today. <http://www.arthritis.org/faqs-about-oa-2.php>. Accessed January 18, 2010.

7. Chitnavis, J., Sinsheimer, J. S., Clipsham, K., Loughlin, et al. (1997). Genetic influences in end-stage osteoarthritis: Sibling risks of hip and knee replacement for idiopathic osteoarthritis. Journal of Bone and Joint Surgery (Br.), 79(4): 660–664.

8. Hunter, D. J., & Lo, G. H. (2008). The management of osteoarthritis: An overview and call to appropriate conservative treatment. Rheumatic Disease Clinics of North America, 34(3): 689–712.

9. Theodosakis, J., Adderly, B., & Fox, B. (1997). The Arthritis Cure. New York: St. Martin's Griffin.

10. National Center for Health Statistics (August 1, 2009). Arthritis Related Statistics. <http://www.cdc.gov/arthritis/data_statistics/arthritis_related_stats.htm>. Accessed January 21, 2010.

11. Pennsylvania Health Care Cost Containment Council (PHC4). (2005). Total Hip and Knee Replacements: Fiscal Year 2002: July 1, 2001, to June 30, 2002. <http://www.phc4.org/reports/hipknee/02/docs/hipkneeFY2002report.pdf>. Accessed January 18, 2010.

12. Bradley, J. D., et al. (1991). Comparison of an anti-inflammatory dose of ibuprofen, an analgesic dose of ibuprofen, and acetaminophen in the treatment of patients with osteoarthritis of the knee. New England Journal of Medicine, 325: 87–91. Cited in Theodosakis et al. (1997), p. 122.

13. Singh, G. (1998). Recent considerations in nonsteroidal anti-inflammatory drug gastropathy. American Journal of Medicine, 105(1), 31S–38S. Cited in Theodosakis et al. (1997), p. 120.

14. Wolfe, M., Lichtenstein, D., & Singh, G. (1999). Gastrointestinal toxicity of nonsteroidal anti-inflammatory drugs. New England Journal of Medicine, 340(24): 1888–1889. Cited in Theodosakis et al. (1997), p. 120.

15. Gross, K. D., & Hillstrom, H. J. (2008). Noninvasive devices targeting the mechanics of osteoarthritis. Rheumatic Disease Clinics of North America 34(3): 755–776.

16. Kirkley, A., Birmingham, T. B., Litchfield, R. B., et al. (2008). A randomized trial of arthroscopic surgery for osteoarthritis of the knee. New England Journal of Medicine, 359(11): 1097–1107.

17. Richmond, J. C. (2008). Surgery for osteoarthritis of the knee. Rheumatic Disease Clinics of North America, 34(3): 815–825.

18. Robbins, S., Waked, E., Allard, P., et al. (1997). Foot position awareness in younger and older men: The influence of footwear sole properties. Journal of the American Geriatric Society, 45(1), 61–66. Cited in Gross & Hillstrom, p. 755.

19. Sharma, L., Song, J., Felson, D. T., et al. (2001). The role of knee alignment in disease progression and functional decline in knee osteoarthritis. Journal of the American Medical Association, 286(2): 188–195.

20. Cluett, J. (n.d.). Zimmer Suspends Marketing of Durom Cup Hip Replace-

ment System amid Reports of Defects, Ceramic Hip Replacement Surgery. <http://www.About.com>

CHAPTER 2

1. Adesman, A. (2009). Baby Facts: The Truth about Your Childs' Health from Newborn through Preschool. Hoboken, NJ: John Wiley.

2. Hunter, D. J., & Lo, G. H. (2008). The management of osteoarthritis: an overview and call to appropriate conservative treatment. Rheumatic Disease Clinics of North America, 34(3): 689–712.

3. Pack, L. G. (1987). Limb salvage. In E. D. McGlammary, Ed., Fundamentals of Foot Surgery, pp. 457–464. New York: Williams and Wilkens.

4. Pack, L. G., & Guastella, G. (1987). Medical and surgical considerations for the patient with mellitus. In E. D. McGlammary, Ed., Fundamentals of Foot Surgery, pp. 376–382. New York: Williams and Wilkens.

5. Pack, L. G., & Lockson, S. S. (1981). Pedal infectons in the vascularly compromised: Evaluation and treatment. Journal of the American Podiatric Association, 71(1): 24–26.

6. Pack, L. G. (1976). Acute adult inflammatory pedal arthritis: A practical guide to diagnosis. Journal of the American Podiatric Medical Association, 66(9): 663–668.

7. Pack, L. G. (1978). Podiatric rheumatology: A guide to diagnosis and treatment. In T. H. Clarke, Ed., Yearbook of Podiatric Medicine and Surgery/ Rheumatology 1978/79, pp. 309–315. Mt. Kisco, NY: Futura.

8. Pack, L. G., Ed. (1979). Rheumatology. In T. H. Clarke, Ed., Yearbook of Podiatric Medicine and Surgery/Rheumatology 1979/80. Mt. Kisco, NY: Futura.

9. Pack, L. G., & Fields, L. S. (1979). Review of the literature. In T. H. Clarke, Ed., Yearbook of Podiatric Medicine and Surgery/Rheumatology 1979/80. Mt. Kisco, NY: Futura.

10. Anandarajah, A. P., Thiele, R. G., & Monu, J. (2009). Erosive Arthritis Is a Progressive Disease That Commonly Affects the Feet. Presentation at the American College of Rheumatology annual meeting, Philadelphia, PA.

11. Niu, J., Felson, D. T., Nevitt, M., Aliabadi, P., et al. (2009). Pain in One

Knee Increases the Risk of Tibiofemoral Osteoarthritis in the Contralateral Knee: The MOST Study. Presentation at the American College of Rheumatology annual meeting, Philadelphia, PA.

12. Mastri, B. (2007). Varus alignment was associated with an increased risk of osteoarthritis of the knee. Journal of Bone and Joint Surgery (Am.), 89(11): 2557.

13. Felson, D. T., Conference chair. (October 2000). Osteoarthritis: New insights. Annals of Internal Medicine, 133(8): 635–646. <http://www.annals.org>. Accessed September 9, 2009.

CHAPTER 3

1. Miller, J. T., Rahimi, S. Y., & Lee, M. (2005). History of infection control and its contributions to the development and success of brain tumor operations. Neurosurgical Focus, 18(4): 1–5.

2. Wikipedia. (n.d.). Ignaz Semmelweis. <http://en.wikipedia.org/wiki/Ignaz_Semmelweis>. Accessed January 21, 2010.

3. Simon, S., Radin, E. L., Paul, I. L., et al. (1972). The response of joints to impact loading. II. In vivo behavior of subchondral bone. Journal of Biomechanics, 5(3): 267–272. Cited in Gross & Hillstrom (2008), p. 767.

4. Radin, E., Parker, H. G., Pugh, J. W., et al. (1973). Response of joints to impact loading. 3. Relationship between trabecular microfractures and cartilage degeneration. Journal of Biomechanics, 6(1): 51–57. Cited in Gross & Hillstrom (2008), p. 767.

5. Radin, E. L., Ehrlich, M. G., Chernack, R., et al. (1978). Effect of repetitive impulsive loading on the knee joints of rabbits. Clinical Orthopaedics and Related Research, 131: 288–293. Cited in Gross & Hillstrom (2008), p. 767.

6. Radin, E. L., Orr, R. B., Kelman, J. L., et al. (1982). Effect of prolonged walking on concrete on the knees of sheep. Journal of Biomechanics, 15(7): 4887–492. Cited in Gross & Hillstrom (2008), p. 767.

7. Burr, D. B., & Radin, E. L. (2003). Microfractures and microcracks in subchondral bone: Are they relevant to osteoarthrosis? Rheumatic Disease Clinics of North America, 29(4): 675–685. Cited in Gross & Hillstrom (2008), p. 767.

8. Gross, K. D., & Hillstrom, H. J. (2008). Noninvasive devices targeting the mechanics of osteoarthritis. Rheumatic Disease Clinics of North America, 34(3): 755–776.

9. Tanamas, S., Hanna, F. S., Cicuttini, F. M., et al. (2009). Does knee malalignment increase the risk of development and progression of knee osteoarthiris? A systematic review. Arthritis and Rheumatism, 61(4): 459–467.

10. Kuhn, D. R., Shibley, N. J., Austin, W. M., & Yochum, T. R. (1999). Radiographic evaluation of weight-bearing orthotics and their effect on flexible pes planus. Journal of Manipulative and Physiological Therapeutics, 22(4): 221–226.

11. MacLean, C. L., Davis, I. S., & Hamill, J. (2008). Short- and long-term influences of a custom foot orthotic intervention on lower extremity dynamics. Clinical Journal of Sports Medicine, 18(4): 338–343.

12. Thorp, L. E., Sumner, D. R., Block, J. A., Moisio, K.C., et al. (2006). Knee joint loading differs in individuals with mild compared with moderate medial knee osteoarthritis. Arthritis and Rheumatism, 54(12): 3842–3849.

13. Reilly, K., Barker, K., Shamley, D., Newman, M., et al. (2009). The role of foot and ankle assessment of patients with lower limb osteoarthritis. Physiotherapy, 95(3): 164–169.

14. Teichtahl, A., A. Wluka, & F. M. Cicuttini (2003). Abnormal biomechanics: A precursor or result of knee osteoarthritis? British Journal of Sports Medicine 37(4): 289–290.

15. Teichtahl, A. J., Morris, M. E., Wluka, A. E., et al. (2006). Foot rotation: A potential target to modify the knee adduction moment. Journal of Sports Science and Medicine, 9(1–2): 67–71.

16. Janakiramanan, N., Teichtahl, A. J., Wluka, A. E., et al. (2008). Static knee alignment is associated with the risk of unicompartmental knee cartilage defects. Journal of Orthopaedic Research, 26(2): 225–230.

17. Sharma, L., Song, J., Felson, D. T., et al. (2001). The role of knee alignment in disease progression and functional decline in knee osteoarthritis. Journal of the American Medical Association, 286(2): 188–195.

18. Khan, F. A., Koff, M. F., Noiseux, N. O., et al. (2008). Effect of local alignment on compartmental patterns of knee osteoarthritis. Journal of Bone and Joint Surgery (Am.), 90(9): 1961–1969.

19. Cicuttini, F., Wluka, A., Hankin, J., & Wang, Y. (2004). Longitudinal study of the relationship between knee angle and tibiofemoral cartilage volume in subjects with knee osteoarthritis. Rheumatology (Oxford), 43(3): 321–324.

20. Theodosakis, J., Adderly, B., & Fox, B. (1997). The Arthritis Cure. New York: St. Martin's Griffin.

21. Hunter, D. J. (2008). Preface. Rheumatic Disease Clinics of North America, 34(3): xiii–xvi.

22. Wilson, D. R., McWalter, E. J., & Johnston, J. D. (2008). The measurement of joint mechanics and their role in osteoarthritis genesis and progression. Rheumatic Disease Clinics of North America, 34(3): 605–622.

23. Zhang, Y., & Jordan, J. M. (2008). Epidemiology of osteoarthritis. Rheumatic Disease Clinics of North America, 34(3): 515–529.

24. White, P. (2009). Update on the AF Public Health Strategic Initiatives and Activities. Presentation to the CDC State Arthritis Program Meeting, July 22.

CHAPTER 4

1. Answers.com. (n.d.). How Many Times Does a Heart Beat in a Lifetime? <nttp://wiki.answers.com/Q/>. Accessed February 5, 2010.

2. Answers.com. (n.d.). How Many Breaths Do Humans Take in an Average Lifetime? <http://wiki.answers.com/Q/>. Accessed January 22, 2010.

3. Georgia Podiatric Medical Association. (n.d.). Foot Facts. <http://www.gapma.com/FootFacts.htm>. Accessed January 26, 2010.

4. Kettering Podiatry Associates. (n.d.) Fun Foot Facts. <http://www.sick-foot.com/footfacts.html>. Accessed January 28, 2010.

5. American Podiatric Medical Association (APMA). (2010). 75,000-Mile Checkup. <http://www.apma.org/MainMenu/News/Campaigns/>. Accessed February 2, 2010.

6. Steinberg, M. D. (2010). Personal communication.

CHAPTER 5

1. Merriam-Webster's Collegiate Dictionary (2006). 11th edn. Springfield, MA: Merriam-Webster.

CHAPTER 6

1. Brady, R. J., J. B. Dean, T. M. Skinner, et al. (2003). Limb length inequality: Clinical implications for assessment and intervention. Journal of Orthopaedic and Sports Physical Therapy 33(5): 221–234.

2. Golightly, Y. M., Allen, K. D., Renner, J. B., et al. (2007). Relationship of limb length inequality with radiographic knee and hip osteoarthritis. Osteoarthritis Cartilage 2007, 15(7): 824–829. Cited in Zhang & Jordan (2008), p. 524.

3. Gross, M. T. (1995). Lower quarter screening for skeletal malalignment: Suggestions for orthotics and shoewear. Journal of Orthopaedic and Sports Physical Therapy, 21(6): 389–404.

4. Song, K. M., Halliday, S. E., & Little, D. G. (1999). Investigation performed at Texas Scottish Rite Hospital for Children, Dallas, TX. Journal of Bone and Joint Surgery, 81: 529–534.

5. Gurney, B., Mermier, C., Rogers, R., Gibson, A., & Rivero, D. (2001). Effects of limb-length discrepancy on gait economy and lower-extremity muscle activity in older adults. Journal of Bone and Joint Surgery, 83: 907–915.

6. Konyves, A., & Bannister, G. C. (2005). Importance of leg length discrepancy after total hip arthroplasty. Journal of Bone and Joint Surgery (Br.), 87(2): 155–157.

7. Khan, H., Fleming, P., & McElwain, J. (2004). Limb length discrepancy following total hip replacement: Incidence and causes. Journal of Bone and Joint Surgery (Br.), Orthopaedic Proceedings, 86-B: 125.

8. Jasty, M., Webster, W., & Harris, W. (1996). Management of limb length inequality during total hip replacement. Clinical Orthopaedics, 333: 165–171.

9. Mayo Clinic. (2008). Statins: Are These Cholesterol-Lowering Drugs Right for You? <http://mayoclinic.com/health/statins/CL00010>. Accessed February 7, 2010.

10. Hill, R. S. (1995). Ankle equinus: Prevalence and linkage to common foot pathology. Journal of the American Podiatric Medical Association, 85(6): 295–300.

11. DiGiovanni, C. W., Kuo, R., Tejwani, N., Price, R., et al. (2002). Isolated gastrocnemius tightness. Journal of Bone and Joint Surgery, 84: 962–970.

12. Lavery, L. A., Armstrong, D. G., & Boulton, A. J. M. (2005). Plantar pressure in a large population with diabetes mellitus. Journal of the American Podiatric Medical Association, 95(5): 464–468.

13. Barrett, S. L., & Jarvis, J. (2005). Equinus deformity as a factor in forefoot nerve entrapment treatment with endoscopic gastrocnemius recession. Journal of the American Podiatric Medical Association, 95(5): 464–468.

14. Szames, S. E., Forman, W. M., Oster, J., Eleff, J. C., & Woodward, P. (1990). Sever's disease and its relationship to equinus: A statistical analysis. Clinical Podiatric Medicine and Surgery, 7(2): 377–384.

15. Meszaros, A., & Caudell, G. (2007). The surgical management of equinus in the adult acquired flatfoot. Clinics in Podiatric Medicine and Surgery, 24(4), 667–685.

16. Higginson, J. S., Zajac, F. E., Neptune, R. R., Kautz, S. A., et al. (2006). Effect of equinus foot placement and intrinsic muscle response on knee extension during stance. Gait and Posture, 23: 32–36.

17. McMulkin, M. L., Barr, K. M., Goodman, M. J., Menown, J. L., et al. (n.d.). Secondary gait compensation in normal adults with an imposed unilateral ankle equinus contracture. Spokane, WA: Motion Analysis Laboratory, Shriners Hospital for Children; Department of Physical Therapy, Washington University.

18. Messier, S. P. (2008). Obesity and osteoarthritis: Disease genesis and nonpharmacologic weight management. Rheumatic Disease Clinics of North America, 34(3): 713–729.

CHAPTER 7

1. Kettering Podiatry Associates. (n.d.). Fun Foot Facts. <http://www.sickfoot.com/footfacts.html>. Accessed January 28, 2010.

2. Steinberg, M. D. (1970). Personal communication.

3. Wikipedia. (n.d.). Arch support. <http://en.wikipedia.org/wiki/Arch_support>. Accessed February 7, 2010.

4. Gélis, A., Coudeyre, E., Aboukrat, P., Cros, P., et al. (2005). Feet insoles and knee osteoarthritis: Evaluation of biomechanical and clinical effects from the literature. Annales de Réadaptation et de Médicine Physique, 48(9): 682–689.

5. Teichtahl, A., A. Wluka, & F. M. Cicuttini (2003). Abnormal biomechanics: A precursor or result of knee osteoarthritis? British Journal of Sports Medicine 37(4): 289–290.

6. White, P. (2009). Update on the AF Public Health Strategic Initiatives and Activities. Presentation to the CDC State Arthritis Program Meeting, July 22.

7. Gross, K. D., & Hillstrom, H. J. (2008). Noninvasive devices targeting the mechanics of osteoarthritis. Rheumatic Disease Clinics of North America, 34(3): 755–776.

8. Williams III, D. S., McClay, D. I., & Baitch, S. P. (2003). Effect of inverted orthoses on lower-extremity mechanics in runners. Medicine and Science in Sports and Exercise, 35(12): 2060–2068.

9. MacLean, C., Davis, I. M., & Hamill, J. (2006). Influence of a custom foot orthotic intervention on lower extremity dynamics in healthy runners. Clinical Biomechanics, 21(6): 623–630.

10. Mündermann, A., Nigg, B. M., Humble, R. N., & Stefanyshyn, D. J. (2003). Foot orthotics affect lower extremity kinematics and kinetics during running. Clinical Biomechanics, 18(3): 254–262.

11. Kerrigan, D. C., Lelas, J. L., Goggins, J., Merriman, G. J., et al. (2002). Effectiveness of a lateral-wedge insole on knee varus torque in patients with knee osteoarthritis. Archives of Physical Medicine and Rehabilitation, 83(7): 889–893.

12. Schmaltz, T., Blumentritt, S., Drewitz, H., & Freslier, M. (2006). The influence of sole wedges on frontal plane knee kinetics, in isolation and in combination with representative rigid and semi-rigid ankle-foot orthoses. Clinical Biomechanics, 21(6): 631–639.

13. Gross, M. T., & Foxworth, J. L. (2003). The role of foot orthoses as an in-

tervention for patellofemoral pain. Journal of Orthopaedic and Sports Physical Therapy, 33(11): 661–670.

14. Marks, R., & Penton, L. (2004). Are foot orthotics efficacious for treating painful medial compartment knee osteoarthritis? A review of the literature. International Journal of Clinical Practice, 58(1): 49–57.

15. Krohn, K. (2005). Footwear alterations and bracing as treatments for knee osteoarthritis. Current Opinion in Rheumatology, 17(5): 653–656.

16. Saxena, A., & Haddad, J. (2003). The effect of foot orthoses on patellofemoral pain syndrome. Journal of the American Podiatric Medical Association, 93(4): 264–271.

17. Rubin, R., & Menz, H. B. (2005). Use of laterally wedged custom foot orthoses to reduce pain associated with medial knee osteoarthritis: A preliminary investigation. Journal of the American Podiatric Medical Association, 95(4): 347–352.

18. Johnston, L. B., & Gross, M. T. (2004). Effects of foot orothoses on quality of life for individuals with patellofemoral pain syndrome. Journal of Orthopaedic and Sports Physical Therapy, 34(8): 440–448.

19. Keating, E. M., Faris, P. M., Ritter, M. A., & Kane, J. (1993). Use of lateral heel and sole wedges in the treatment of medial osteoarthritis of the knee. Orthopedic Reviews, 22(8): 921–924.

20. Dananberg, H. J., & Guiliano, M. (1999). Chronic low-back pain and its response to custom-made foot orthoses. Journal of the American Podiatric Medical Association, 89(3): 109–117.

21. Powell, M., Seid, M., & Szer, I. S. (2005). Efficacy of custom foot orthotics in improving pain and functional status in children with juvenile idiopathic arthritis: A randomized trial. Journal of Rheumatology, 32(5): 943–950.

22. de P. Magalhäes, E., Davitt, M., Filho, D. J., Battistella, L. R., et al. (2006). The effect of foot orthoses in rheumatoid arthritis. Rheumatology (Oxford), 45(4): 449–453.

23. Woodburn, J., Barker, S., & Helliwell, P. S. (2002). A randomized controlled trial of foot orthoses in rheumatoid arthritis. Journal of Rheumatology, 29(7): 1377–1383.

24. Woodburn, J., Helliwell, P. S., & Barker, S. (2003). Changes in 3D joint kinematics support the continuous use of orthoses in the management of painful rearfoot deformity in rheumatoid arthritis. Journal of Rheumatology, 30(11): 2356–2364.

25. Li, C. Y., Imaishi, K., Shiba, N., Tagawa, Y., et al. (2000). Biomechanical evaluation of foot pressure and loading force during gait in rheumatoid arthritic patients with and without foot orthosis. Kurume Medical Journal, 47(3): 211–217.

26. Chalmers, A. C., Busby, C., Goyert, J., Porter, B., et al. (2000). Metatarsalgia and rheumatoid arthritis: A randomized, single blind, sequential trial comparing 2 types of foot orthoses and supportive shoes. Journal of Rheumatology, 27(7): 1643–1647.

27. Hodge, M. C., Bach, T. M., & Carter, G. M. (1999). Orthotic management of plantar pressure and pain in rheumatoid arthritis. Clinical Biomechanics, 14(8): 567–575.

28. Clark, H., Rome, K., Plant, M., O'Hare, K., et al. (2006). A critical review of foot orthoses in the rheumatoid arthritic foot. Rheumatology (Oxford), 45(2): 139–145.

29. Mejjad, O., Vittecoq, O., Pouplin, S., Grassin-Delyle, L., et al. (2004). Foot orthotics decrease pain but do not improve gait in rheumatoid arthritis patients. Joint Bone Spine, 71(6): 542–545.

30. Slattery, M., & Tinley, P. (2001). The efficacy of functional foot orthoses in the control of pain in ankle joint disintegration in hemophilia. Journal of the American Podiatric Medical Association, 91(5): 240–244.

31. Brouwer, R. W., Jakma, T. S., Verhagen, A. P., Verhaar, J. A., et al. (2005). Braces and orthoses for treating osteoarthritis of the knee. Cochrane Database of Systematic Reviews, January 25(1): CD004020.

CHAPTER 8

1. Merriam-Webster's Collegiate Dictionary (2006). 11th edn. Springfield, MA: Merriam-Webster.